AF557764

Plant Breeding: Scientific Methodologies and Technologies

Plant Breeding: Scientific Methodologies and Technologies

Dr. Prashant Kumar Sirohi

Plant Breeding: Scientific Methodologies and Technologies

ISBN 978-93-5111-592-2

Published in 2015 in India by

RANDOM PUBLICATIONS

4376-A/4B, Gali Murari Lal, Ansari Road
New Delhi-110 002
Phone : +9111-43580356, 011-23289044, 011-43142548
e-mail: sales@randompublications.com,
info@randompublications.com, randomexports@gmail.com

Reprinted 2023

Type Setting by : Friends Media, Delhi-110089
Digitally Printed at : Replika Press Pvt. Ltd.

Preface

Plant breeding is the purposeful manipulation of plant species in order to create desired genotypes and phenotypes for specific purposes. This manipulation involves either controlled pollination, genetic engineering, or both, followed by artificial selection of progeny. Plant breeding often, but not always, leads to plant domestication. Plant breeding has been practiced for thousands of years, since near the beginning of human civilization. It is now practiced worldwide by government institutions and commercial enterprises. International development agencies believe that breeding new crops is important for ensuring food security and developing practices through the development of crops suitable for their environment. Classical plant breeding uses deliberate interbreeding (crossing) of closely or distantly related individuals to produce new crop varieties or lines with desirable properties. Plants are crossbred to introduce traits/genes from one variety or line into a new genetic background. For example, a mildew-resistant pea may be crossed with a high-yielding but susceptible pea, the goal of the cross being to introduce mildew resistance without losing the high-yield characteristics. Progeny from the cross would then be crossed with the high-yielding parent to ensure that the progeny were most like the high-yielding parent, (backcrossing). The progeny from that cross would then be tested for yield and mildew resistance and high-yielding resistant plants would be further developed. Plants may also be crossed with themselves to produce inbred varieties for breeding.

I would like to thank my team for standing beside me throughout my career and writing this book. My special thanks go to "Random Publications" who have published the book.

– Dr. Prashant Kumar Sirohi

Contents

1

The Plant Kingdom

Since the time of Aristotle, biologists have sought to classify organisms. At first the purpose was ease of identification ("artificial" classification schemes). Carolus Linnaeus (1707–1778), arguably the greatest of the pre-modern Naturalists, sought to classify plants and other organisms according to affinity groups that reflected the mind of the Creator.

Later, after Darwin, the goal of classification was to show evolutionary relationships ("natural" classification schemes). For the past 150 years, biologists have emphasized natural systems of classification and have attempted to define morphological criteria that reveal evolutionary relationships.

We now know that morphology, the form and structure of organisms, is the end product of the actions of genes. Virtually all of the information needed to form a complete organism is encoded in its DNA sequences, both nuclear and cytoplasmic (mitochondria and chloroplasts). DNA sequence analysis has thus provided evolutionary biologists with a powerful new tool for arriving at a truly natural classification system

On the basis of phylogenetic analyses of highly conserved DNA sequences, living organisms have been divided into three major domains: Bacteria, Archaea, and Eucarya.

- The common ancestor of all the organisms first gave rise to the Bacteria and the common ancestor of the Archaea and the Eucarya.
- The Archaea branch off from the Eucarya lineage.
- The Eucarya common ancestor acquires mitochondrial endosymbiont (an alpha-proteobacterium-like cell).
- A heterogeneous group of eukaryotes called protists branch off the lineage leading to plants, fungi, and animals.
- The common ancestor of fungi and animals form a branch, followed by a divergence into the fungal and animal lineages.
- The common ancestor of plants, green algae, red algae, and glaucophytes acquires chloroplast endosymbiont (a cyanobacterium).
- The three lineages of glaucophytes, red algae, and green algae diverge.

- Various lineages of protists acquire chloroplasts via green or red algal endosymbionts.
- The earliest branch of green algae (Green Algae I) diverge.
- The later branch of green algae (Green Algae II, Characeae, Coleochaetales) diverge.
- The remaining lineage leads to plants.

Bryophytes are small (rarely more than 4 cm in height), very simple land plants, and the least abundant in terms of number of species and overall population. Bryophytes do not appear to be in the direct line of evolution leading to the vascular plants; rather, they seem to constitute a separate minor branch. Bryophytes include mosses, liverworts, and hornworts. These small plants have life cycles that depend on water during the sexual phase. Water facilitates fertilization, the fusion of gametes to produce a diploid zygote, a feature also seen in the algal precursors of these plants. Bryophytes are like algae in other respects as well: They have neither true roots nor true leaves, they lack a vascular system, and they produce no hard tissues for structural support. The absence of these structures that are important for growth on land greatly restricts the potential size of bryophytes, which, unlike algae, are terrestrial rather than aquatic.

The ferns represent the largest group of spore-bearing vascular plants. In contrast to the bryophytes, ferns have true roots, leaves, and vascular tissues, and they produce hard tissues for support. These architectural features enable ferns to grow to the size of small trees. Although ferns are better adapted to the drying conditions of terrestrial life than bryophytes are, they still depend on water as a medium for the movement of sperm to the egg. This dependence on water during a critical stage of their life cycle restricts the ecological range of ferns to relatively moist habitats.

The most successful terrestrial plants are the seed plants. Seed plants have been able to adapt to an extraordinary range of habitats. The embryo, protected and nourished inside the seed, is able to survive in a dormant state during unfavourable growing conditions such as drought. Seed dispersal also facilitates the dissemination of the embryos away from the parent plant.

Another important feature of seed plants is their mode of fertilization. Fertilization in seed plants is brought about by wind- or insect-mediated transfer of pollen, the gamete-producing structure of the male, to the sexual structure of the female, the pistil. Pollination is independent of external water, a distinct advantage in terrestrial environments. Many seed plants produce massive amounts of woody tissues, which enable them to grow to extraordinary heights. These features of seed plants have contributed to their success and account for their wide range. There are two categories of seed plants: gymnosperms (from the Greek for "naked seed") and angiosperms (based on the Greek for "vessel seed," or seeds contained in a vessel). Gymnosperms are the less advanced type; about 700 species of gymnosperms are known. The largest group of gymnosperms is the conifers (“cone-bearers”), which include such commercially important forest trees as pine, fir, spruce, and redwood.

Two types of cones are present: male cones, which produce pollen, and female cones, which bear ovules. The ovules are located on the surfaces of specialized structures called cone scales. After wind-mediated pollination, the sperm reaches the egg via a pollen tube, and the fertilized egg develops into an embryo. Upon maturation, the cone scales, which are appressed during early development, separate from each other, allowing the naked seeds to fall to the ground.

Angiosperms, the more advanced type of seed plant, first became abundant during the Cretaceous period, about 100 million years ago. Today, they dominate the landscape, easily outcompeting their cousins, the gymnosperms. About 250,000 species are known, but many more remain to be characterized. A typical angiosperm life cycle, that of *Zea mays* (corn). The major innovation of the angiosperms is the flower; hence they are referred to as *flowering plants.* There are other anatomical differences between angiosperms and gymnosperms, but none so crucial and far-reaching as the mode of reproduction.

CELL - STRUCTURE

CELL

Cell is the unit of structure and function. They are the building blocks of an organism. Irrespective of the nature of organisms (plant or animal) they are either made up of single cell or many cells, the former is called unicellular and later is called multicellular organisms; in the later cells are differentiated into various kinds and they are grouped into tissues, which perform special and special function.

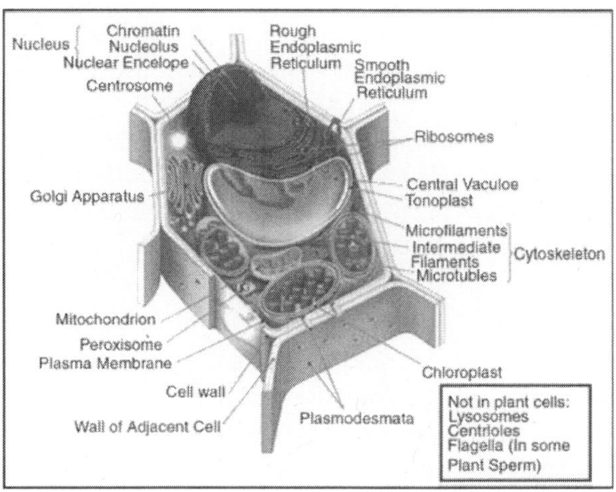

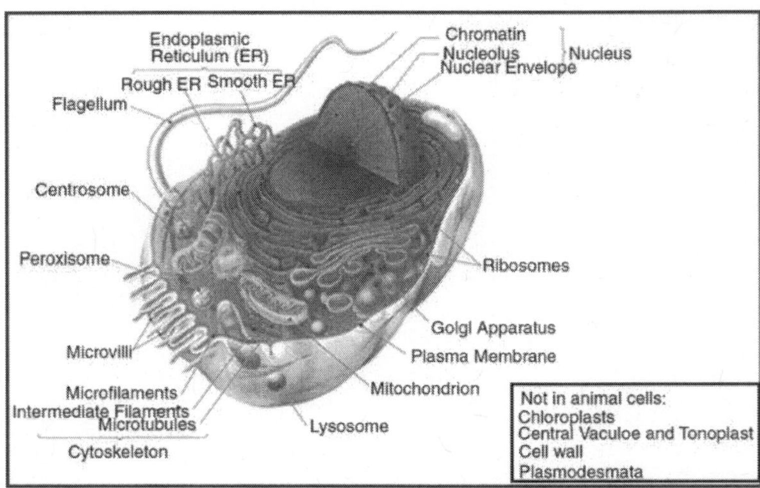

Example meristematic cells perform repeated cell divisions, phloem cells- conduct food material, sclerenchyma- mechanical, xylem- conduction of water and mineral salts and so on. Nevertheless, all these different types of cells are derived from the same embryonic cells.

The development of various cell types from a single cell is determined and regulated by a process called differentiation, which in turn is controlled by differential concentrations of plant hormones.

Added to this, organ differentiation is another fascinating aspect of development. All these processes are regulated by differential gene regulation in response to environmental stimulus and phytohormones.

CELLULAR COMPOSITION

All cells are made up of a semi viscous fluid called protoplasm, which is considered as the physical basis of life, for it controls all biochemical reactions of the cell. In fact it is the microcosm of life with many secrets not known to us. The protoplasm is colloidal in nature, because many cell colloidal sized structures and macromolecules are suspended in it. It also exhibits sol and gel properties.

The granular nature of the protoplasm is due to the presence of many tiny organelles. Vacuoles of various types are also found, but in plant cells when it is matured, a large central vacuole is present and it is separated from the rest of the protoplasm by a single unit membrane called tonoplast. The fluid present is called cell sap.

There are many common features between plant and animal cells; the former is distinguished by the presence of distinct cell wall and plastids, which are totally absent in animal cells. However centrosomes are invariably present in animal cells and rarely in plant cells with exception of some lower plant unicellular algae like Chlamydomonas.

CHEMICAL COMPOSITION

When cells are subjected to chemical analysis the following compounds are found (approximate value).

Compound	Animal	Plant Cell
Carbohydrates	20%	30%
Proteins	45%	40%
Lipids	30%	25%
RNA	1%	0.4%
DNA	0.2%	0.4%
Inorganic and	others 3.8%	3.6%

- *Carbohydrates*: These are organic compounds consisting of C, H and O. The basic structural components of carbohydrates are monosaccharide sugars consisting of 3 to 7 carbons. Example: Glucose (6c), Fructose (6c), Erythrose (4c), Xylose (5c), etc. such monomers by undergoing polymerization develop into long chained polysaccharides, such as cellulose, cellobioses, starch and glycogen (animal starch). Cellulose is made up of glucose units linked by b1>4 linkages and it is an important component of cell wall. Similarly starch is also a polymer of á1-4 linked glucose units and it is the main source of energy for all living cells.
- *Proteins*: Proteins are the most important organic components of cells, for they act as structural as well as functional molecules. With out proteins life cannot exist, though DNA is the genetic material it is proteins that make it or break it; DNA is master library with information, that cannot be changed. They are made up of basic building blocks called amino acids. The polymers of amino acid residues are called polypeptides (proteins) which exhibit different structural conformation (shape). The 3-D shapes are specific and characteristic for a particular protein, thus they exhibit specific structure and function. Exemple: contractile proteins (muscles), transport proteins (Hemoglobin), enzyme proteins, hormonal proteins (insulin), antibody proteins IgG etc. Almost all biochemical activities, including growth and development are controlled by proteins, without which cell ceases to live.
- *Lipids*: Fatty acids and their derivatives are very important for two reasons, firstly lipids like lecithin, phosphotidyl ethanolamine, sphingolipids, glycolipids, steroids and others are part of the cellular membranes; thus they contribute to the structure of the membranes. Secondly lipids also act as food reserve and provide energy by oxidation.
- *Nucleic acids*: These are the polymers of nucleotides, consisting of nitrogenous bases, phosphates and pentose sugars. There are two

types of nucleic acid – 1) Deoxy-ribose nucleic acid (DNA), 2) Ribose nucleic acid (RNA). DNA is mostly found in chromosomes, it is the repository of genetic information and it provides information for the synthesis of proteins.

- *Inorganic/Organic factors*: Many inorganic metals Fe, Mg^{2+}, Mn^{2+}, Mo, ca^{2+} etc. are very important, for they are either the integral part of some organic molecules or act as co-factor in enzymatic activities. Certain organic factors like important for cellular metabolism, because many of the vitamins act as co-enzymes; these are required for enzymatic activities, without which cellular processes come to stand still.

SIZE AND SHAPE

The size of cells varies from 10 micron to many centimetres in length. For example cotton fibres are several mm in length The shape ranges from spherical, isodiametric, and hexagonal to tubular. This is genetically predetermined to perform different functions.

When the cell is observed through light microscopes, which may have the maximum resolution of about 2000 times, very few details can be made out. On the other hand, if sections of the cells are observed through electron microscope, which has a resolution power ranging from 50,000 to 150,000 times enlargement, even smaller structures stand out clearly.

Inspite of high resolution, it is not possible to make out all the structural details.

The following structures are found in the cell:

- Cell wall,
- Plasma membrane,
- Nucleus,
- Plastids,
- Mitochondria,
- Golgi complex,
- Ribosomes,
- Cytoskeleton,
- Micro bodies,
- Centrosomes,
- endoplasmic reticulum,
- central vacuole and
- Non living cell inclusions.

Cell Wall

Only plant cells and bacterial cells posses a protective structure as the cell wall outside the plasma membrane. Bacterial cell wall is firmly adpressed to the underlying plasma membrane.

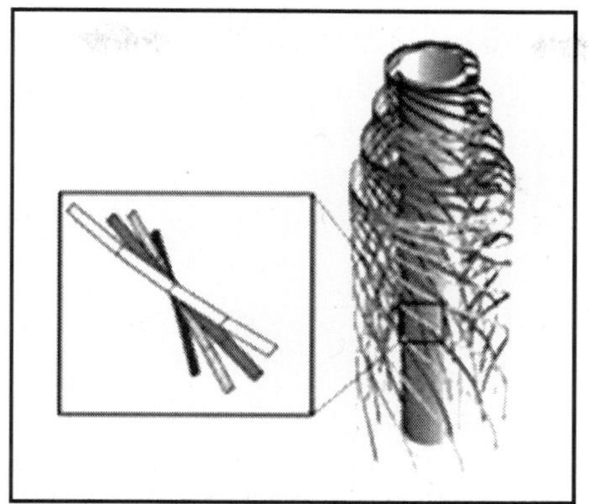

Its chemical composition and structure is more complex. However, it is basically made up of long polymers of glucosamine (NAG) and muramic acids (NAM), which in turn are cross linked by short-pentamer oligopeptides; thus they form a mat like structural layers around the plasma membrane.

Hundreds of such layers are deposited one above the other to form a very tough wall.

Many bacterial cells produce a mucilaginous pectose layers outside the cell wall and this layer is called the capsule. But many bacterial cells do contain another lipid layer studded with proteins, oligosaccharides and glycol and proteoglycans.

But higher plants have cell walls mainly made up of cellulose fibres. Addition to these, pectins, hemicelluloses and lignin are also deposited on primary cellulose layer; the thickening is only at the later stages of development.

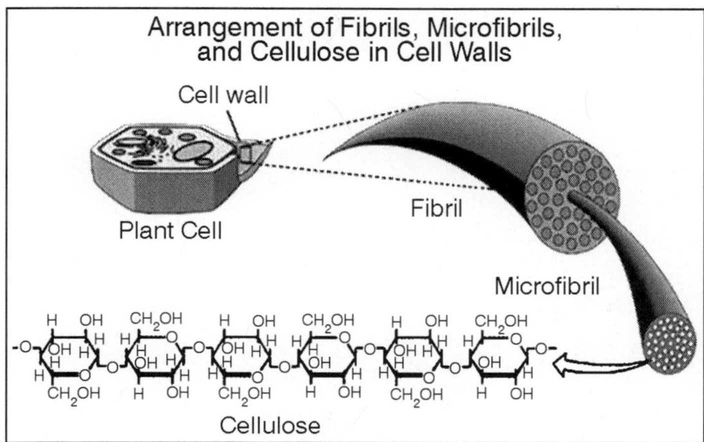

With proper staining, if a group of cells are observed, the cells appear to be held together by a kind of cementing material called middle lamella. It is made up if calcium pectate. Next to calcium pectate layer, the cell wall that is laid later is differentiated into 2 or 3 distinct layers. They are primary wall, secondary wall and tertiary wall.

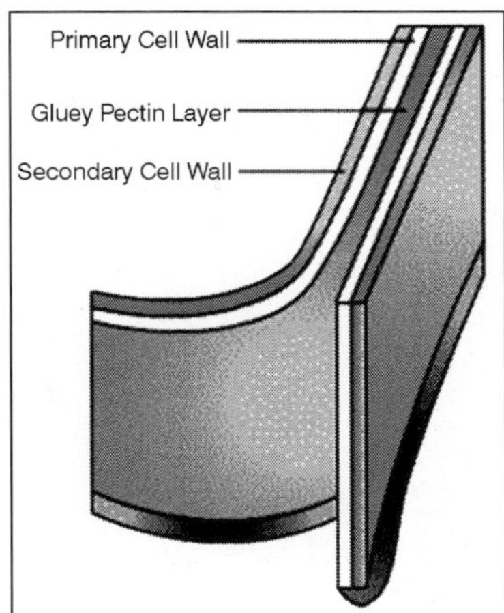

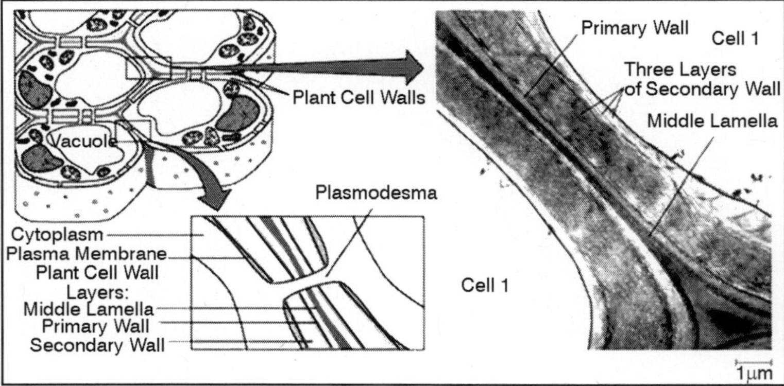

The primary wall is the first true cell wall to be deposited and it is purely made up of cellulose. With the development of the cell, the additional layers are deposited, called secondary and tertiary wall layers, like in the case xylem and sclerenchyma.Certain areas in the cell walls remain without thickening and these are called pit areas, through which fine protoplasmic strands transverse across between two cells and provide a continuum to protoplasm; they are called plasmodesmata.

The primary cell wall, which is the first wall layer to be deposited, is made up of cellulose fibres. These fibres are considerably longer and deposited layers after layers; oriented longitudinally, transversely or obliquely.The deposition and the orientation of these layers are aided by microtubules that are found on the inner face of the plasma membrane.

Each cellulose fibre is made up of 8000 to 120,800 D- Glucose units, which are linked to each other by glycosidic bonds to form a long chain of glucose units, which show helical conformation. Hundreds of such cellulose threads are grouped into a bundle called micelles; these in turn are aggregated into micro fibrils which by further aggregation develop in to macro fibrils.

Such micro/macro fibrils are deposited regularly either longitudinally or transversely to form uniform layers. These fibres are embedded in a matrix made up of pectate substances and hemi cellulose materials like polyxylose and others.

Primary cell wall	*Secondary cell wall*
Primarily cellulose	Few layers of cellulose
Hemicelluloses	Xylose, mannose
Pectic	Complex lignins
Elastic	Non-elastic
Laid on middle lamellae	Laid over primary wall
1-3 micron thick	>5-10 micron thick

CELL MEMBRANES

These are the most important structures of the cell, for they are responsible for protecting and separating the protoplasm from the external environment.It helps in selective uptake and transport of ions, provides surface area for many biochemical signal transduction reaction and also helps in specialized functions. In fact most of the cell organelles are bounded by membranes. Notwithstanding this, a large number of functional molecules are integrated into these membranes.

CHEMICAL COMPOSITION

Almost all membranes are made up of proteins and lipids. The ratio between proteins and lipids may vary in different membranes, but generally it is equal. It is not uncommon to see some carbohydrates as glycoproteins associated with membrane on the outer surface of plasma membranes. Proteins found in membrane are not of the same kind, but differ in their structure, chemical composition and function. Some of them are large and many of them may be smaller. However, most of the proteins are globular in nature having either hydrophilic or hydrophobic or both the characters. Lipids, on the other hand are of various kinds.

The common lipids found in the membranes are phosphotidyl choline (Lecithin), phosphotidyl ethanolamine, Phospholipids glycerol, cholesterol, etc. some of the phospholipids exhibit hydrophilic groups at one end and hydrophobic at the other end. Nevertheless, the composition of lipids varies from membrane to membrane, for they have different functions to perform.

Membrane Structure

The structural organization of various compounds with in the membrane was an enigma for a long time. With the advancement of biological techniques, it has been found that the membranes exhibit fluid-mosaic structure (Singer and Nicholson 1972). Basically lipids form bilayers by organizing hydrophobic layers facing each other and the charged region out side. Proteins, depending upon the charged nature, some are found on the surface and some are embedded. But proteins and lipids of various kinds are oriented towards each other in such a way, they exhibit semi-solid (crystalline) and semi fluid properties. The arrangement of lipids and proteins is of mosaic pattern, where many proteins are half buried in the lipid bilayers and some traverse the entire cross section of the lipid layer in such a way a part it is buried in the core and some are located at the peripheral surface.

The structural and chemical heterogeneity is the hall mark of these membrane structures. This model explains various biological phenomena observed in most of the biological systems.

Various membranes found in the cell, like plasma membrane, endoplasmic reticulum; organelle membranes exhibit basically the same structural pattern, but vary in their lipid and protein composition.

Some of them are single unit membranes (plasma membrane, tonoplast, Lysosomes and peroxisomes) and some have double membrane systems (endoplasmic reticulum, nuclear membrane, golgi membranes chloroplast membrane and mitochondrial membrane). However they show differences in their chemical composition structural features.

Plasma Membrane

This is the outer most membrane of the cell, within which all protoplasmic structures are included. In plants, cell wall acts as an additional protective layer, but in animal cell, plasma membrane itself is the bounding membrane.

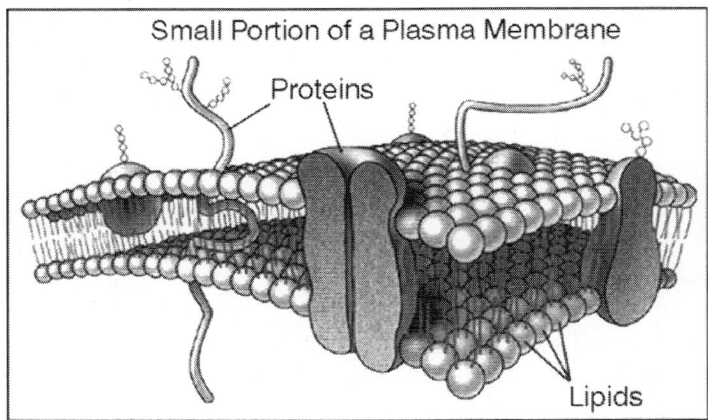

This membrane is in immediate contact with the external environment, and performs various functions like osmosis, selective absorption of various mineral nutrients, accommodates innumerable signal transducing receptors for various stimuli (electrical, light mechanical and chemical) and invariably exhibits dynamic properties. Proteins found in the plasma membrane are vectorial (directional) arranged. At some places, the plasma membrane shows inward projections and they are in continuity with the endoplasmic reticulum. Plasma membrane is also the site for pinocytosis and phagocytosis. Thus the plasma membrane exhibits unique but varied properties of its own. It exhibits dynamic fluidity never to be constant and stagnant.

Endoplasmic Reticulum

It is a labyrinthine net work of double membrane sheets. They are found in all living cells with the exceptions of mature erythrocytes and cells of bacteria. Endoplasmic reticulum (ER) occupies more than 50 to 90% of total cell volume. It is in contact with the outer plasma lemma and the outer nuclear membrane.

ER is made up of two single unit membranes folded to form adpressed sheets, which enclose a channel or cisternae. ER is extensively branched, thus the surface area for reactions is enormously increased.

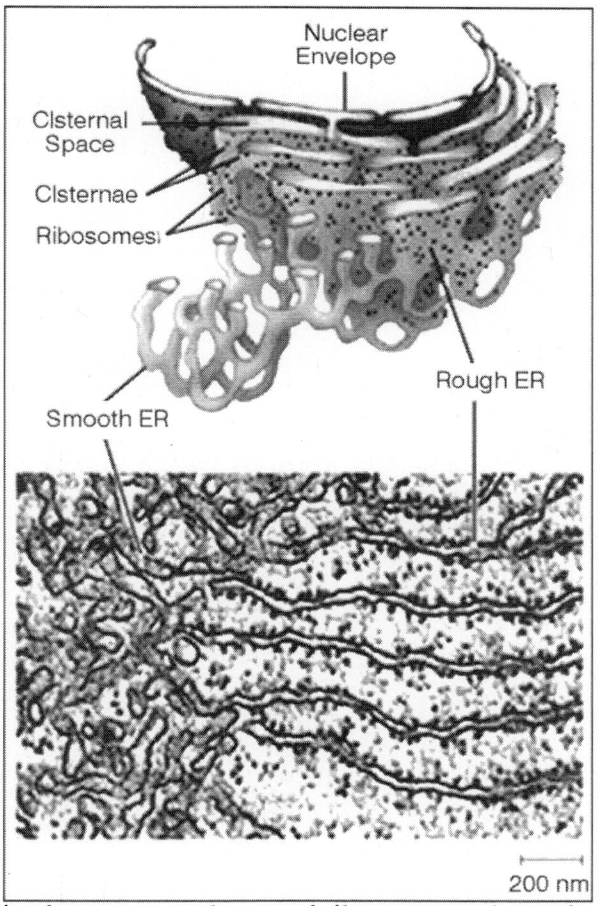

On the basis of presence or absence of ribosomes on the surface, two classes of ER can be recognized:

- Smooth ER (SER) is without ribosomes,
- Rough ER (RER) has innumerable ribosomes on its outer surface.

These membranes are highly mobile and they undergo rapid flux, changing from SER to RER and RER to SER. Added to this, the entire ER exhibit continuous sweeping movements which help in the distribution of cellular components in the matrix. The ER membranes are supported by microtubule skeletal network. These membranes are highly dynamic and show rapid turnover. Functionally ER exhibits various activities like synthesis of proteins, mechanical support for the fluid protoplasm, synthesis and storage of lipids, synthesis and storage and transport of proteins to different destination through golgi membranes. Some of the functions are detoxification, transport of various cellular components, formation of micro bodies, formation of secretory vesicles, cell plate and others. The ER is at the heart of various cellular activities. Furthermore, during the development of nucleus and other cell organelles like, chloroplast, mitochondria, golgi complex, micro bodies, Lysosomes etc. ER provides membrane fractions to them. Exchange of membrane components among them is pervasive and common.

Golgi Bodies

A group of membrane cisternae, discovered by Camillo Golgi in 1890, are called as Golgi bodies. They are present in all cells except bacterial and blue-green algal cells. These structures vary in number from cell type to cell type. In secretory tissues like thyroid and liver they are present in large numbers than in other type of cells. They are abundant in secretory surfaces like stigma of the pistils.

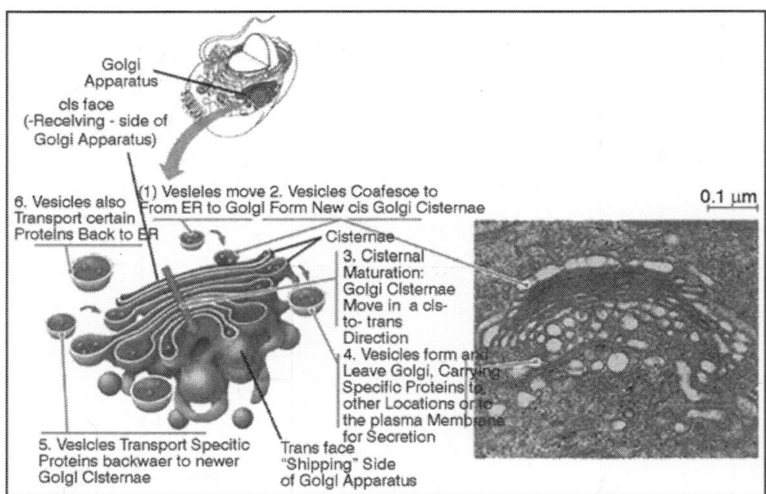

Golgi bodies are made up of a group of stacked membrane cisternae. Some times these membranes structures show extensive reticular network. Generally the distal ends of double membranes are dilated into vesicles and some of the vesicles are in the process of pinching off. An interesting feature is that the Golgi complex is surrounded by ER and at some places they look like in continuity with

ER, but these proximal membranes of ER are free from ribosomes called SER (smooth ER). The stacked Golgi membranes have two faces, *i.e.* formation face is called cis face and maturation face as trans face. The formation face has convex surface and has a number of small vesicles pinched off from SER.

In fact certain proteins synthesized on the RER, enter into the lumen of ER and hence they are transported in the lumen towards the transitional membranes which are in close association with Golgi complex and then the protein containing ER membranes pinch off as vesicles. These in turn fuse with one another or with golgi cis membranes and develop into membranous sacs. Within these membranous sacs, proteins and such products get further modified. Later such products are sorted out and get enclosed and pinched off in the form of vesicles from maturation face called Trans surface of the Golgi complex. Similarly many secretary products that are synthesized on RER and transported into Golgi complex, later the matured products get enclosed in vesicles and budded off. The golgi derived vesicles are loaded with proteins that are specifically targeted to various destinations. Thus, Golgi bodies perform various complex processes, like glocolysation of proteins, synthesis of cell wall polysaccharides, maturation of zymogen granules, formation of primary lysosomes, secretion of lipid bodies, acrosome formation, neural secretion etc. Nevertheless, the participation of ER is every essential in the function of Golgi membranes. Golgi membranes are associated with transport proteins and they are responsible for transportation from one site to the other. In all the above mentioned processes, packing, maturation and secretion of specific substances are the most important events of Golgi functions. The golgi complex is at the heart of membrane flow.

WATER AND PLANT CELLS

The water properties of cohesion, adhesion and surface tension, give rise to the phenomenon of capillarity, the movement of water for small distances up a capillary tube. The smaller the tube radius, the higher the capillary rise. How far the water will move may be calculated using the following formula:

$$\text{Capillary rise} = \frac{14.9 \times 10^{-6}\text{m}^2}{\text{radius}}$$

where both capillary rise and radius are expressed in meters.

For a xylem vessel with 25 μm radius, the capillary rise is about 0.6 m. This distance is much too small to be significant for water transport up tall trees.Fibrous materials such as cell walls can act like wicks to draw water by capillarity from nearby xylem. This capillary action ensures that cell wall surfaces that are directly exposed to the air, such as those in leaf mesophyll, remain wetted and do not dry out. Because the cell wall capillaries have a tiny radius, about 10 m^{-8}, very large physical forces can be generated in the water just below the evaporative surfaces of cell walls.From Fick's law, one can derive an expression for the time it takes for a substance to diffuse a particular distance. If one defines conditions such that all the solute molecules are concentrated at the starting position, then the concentration front moves away from the starting position. As the substance diffuses away from the starting point, the concentration gradient becomes less steep (Δc_s decreases) and thus net movement becomes slower.The time it takes for the substance at any given distance from the starting point to reach one-half of the concentration at the starting point (t_c = ½) is given by the following equation:

$$t_{c=1/2} = \frac{(\text{distance})^2}{D_s} K$$

where K is a constant and D_s is the diffusion coefficient. The above equation shows that the time required for a substance to diffuse a given distance increases in proportion to the *square* of that distance. Let us consider two numerical examples. First, how long it would take a small molecule to diffuse across a typical cell? The diffusion coefficient for a small molecule like glucose is about 10^{-9} m^2 s^{-1}, and the cell length may be 50 μm. Thus, for this example:

$$t_{c=1/2} = \frac{\left(50 \times 10^{-6}\text{m}\right)^2}{10^{-9}\text{m}^2\text{s}^{-1}} = 2.5 \text{ s}$$

This calculation shows that small molecules diffuse over cellular dimensions rapidly. What about diffusion over longer distances? Calculating the time needed for the same substance to move a distance of 1 m (*e.g.*, the length of a corn leaf), we find:

$$t_{c=1/2} = \frac{(1\text{m})^2}{10^{-9}\text{m}^2\text{s}^{-1}} = 10^9 \text{ s} \approx 32 \text{ years}$$

a value that exceeds by orders of magnitude the life span of a corn plant, which lives only a few months. This shows that diffusion in solutions can be effective within cellular dimensions but is far too slow for mass transport over long distances. Diffusion is of great importance as a driving force for the water vapour lost from leaves, because the diffusion coefficient for a molecule in air is much greater than in aqueous solutions.

COMPONENTS OF WATER POTENTIAL

Students planning further study of plant water relations should note that the components of water potential defined in the text are sometimes given different names and symbols. In particular, the equation

$$\Psi_w = \Psi_s = \Psi_p$$

is often replaced by the following equivalent equation:

$$\Psi_w = -\pi + P$$

In this alternative convention, P is the same as Ψ_p. It is the hydrostatic pressure of the solution, and may be positive, as in turgid cells, or negative, as in xylem water. The symbol π is called osmotic pressure and is the negative of Ψ_s. That is, π has positive values, and Ψ_s has negative values. "Osmotic pressure" is the term that physical chemists, zoologists, and many others use to denote the effect of dissolved solutes on the free energy of water. Most handbooks of physics and chemistry use the term "osmotic pressure" and the symbol π. The negative sign in front of π in the equation above accounts for the reduction in water potential (Ψ_w) by dissolved solutes. Thus $\Psi_s = -\pi$. A very interesting, if somewhat unconventional, account of the history and physical meaning of osmotic pressure is given by Hammel and Scholander.

Unfortunately, some authors have mixed the conventions for Ψ_s and π, leading to unnecessary confusion about what is meant by the symbol π. Thus, π is sometimes incorrectly called osmotic potential instead of osmotic pressure, and it may be used either as a positive quantity or as a negative quantity.

Regarding matric potential, it is usually designated by the symbol t.

TURGOR PRESSURES EXIST

One of the challenging aspects of understanding plant water relations is the remarkable range of pressures, both positive and negative, that occur within the bodies of plants. Negative pressures (or tensions), which depend upon the cohesive strength of water coupled with the strength of lignified cell walls to resist deformation, play an important role in water transport through the xylem. Positive pressures, which depend upon the semipermeable nature of the plasma membrane and the elastic nature of primary cell walls, occur in all hydrated living plants cells but can be especially large in sieve tubes and guard cells. Living plant cells are typically assumed to have only positive pressures. However, there appears to be no reason that negative pressures could not also occur within the cytoplasm of living plant cells. This topic explores the existence and potential role of negative turgor pressures in plants.

When a cell loses water in air, turgor declines and solute concentrations increase. At the turgor loss point ($\Psi_p = 0$), the hydrostatic pressure in the cell sap is equal to atmospheric pressure, meaning that no net force is exerted on the cell wall. If water continues to be lost from the cell, the pressure within the cytoplasm drops below atmospheric pressure, resulting in a force imbalance that collapses the cell wall. The deformation of living cells upon desiccation is called cytorrhysis. Note that the plasma membrane is pressed against the cell wall throughout desiccation (*i.e.*, plasmolysis does not occur) because the hydrostatic pressure in cytoplasm remains greater than the hydrostatic pressure in apoplast.

In the living cells described above, a decrease in cell water potential below the turgor loss point is balanced by an increase in solute concentration as the volume of the cell decreases. Thus, true "negative" pressures do not develop. In contrast, xylem conduits have rigid cell walls that resist deformation, allowing them to sustain negative pressures without imploding. This raises the question of what happens when water is lost from living cells that have thick walls or are embedded in a rigid matrix of cells (*e.g.*, xylem parenchyma). Might these cells resist deformation and thus develop negative turgor pressures?

It is important to emphasize that this is a controversial area, due in large part to the absence of direct measurements of negative turgor pressures within cells. However, before reviewing the evidence for negative turgor pressures, it is worthwhile to consider what physiological effects might result from the development of such negative pressures in living cells. The major outcome of negative turgor is that it allows stiff-walled cells to decrease in water potential without undergoing major changes in cell volume or osmotic concentration. Because cytorrhysis might cause physical damage to the wall and/or cell membranes, while the high concentrations of solutes resulting from the reduction in cell volume might adversely affect the conformation of membranes and proteins, it is possible to imagine a physiological role or benefit for the development of negative turgor pressures in plant cells.

We now turn this around and ask if there are any downsides to the generation of negative turgor pressures in living cells? In the xylem, the primary issue associated with negative pressures is cavitation; could this also be a concern for living cells? The primary mechanism for cavitation in plants is air seeding, reflecting the fact that the probability of the de novo formation of gas voids in water (either by homogeneous or heterogeneous nucleation) is extremely low. In air seeding, air is sucked in through the cell wall, where it then "nucleates" the transition to the vapour phase. For air seeding to occur in a living cell experiencing negative pressures, air would have to be pulled through the very small pores of the cell wall. While one can imagine this happening, the movement of air across the cell wall would result in plasmolysis and thus the immediate release of any tension in the cytoplasm because the plasma membrane is not capable of withstanding any significant pressure.

Other costs associated with the ability of living cells to generate negative pressures include the metabolic costs of producing rigid cell walls, as well as any limitations lignified walls might place on physiological function. In addition, a strategy for avoiding cell damage due to desiccation via the generation of negative pressures might impose limitations on cell size. The strength of a cell to withstand collapse (and thus generate negative pressures) is inversely proportional to cell size.

Evidence for negative turgor pressure in plants is limited, in part reflecting the fact that few researchers have devoted much attention to this issue. Living cells with flexible (unlignified) cell walls deform relatively easily upon desiccation and thus do not appear to support negative pressures. However, measurements of the forces needed to collapse cell walls suggest that living cells with thick walls can withstand forces in the range of 1.0 MPa. Visual examination of tissues adapted to withstand very cold temperatures also provides indirect evidence for negative turgor. For example, frozen ray parenchyma cells often do not exhibit significant deformation, despite the very strong desiccatory effects (low water potential) of extracellular ice. One can hypothesize that the existence of negative pressures within these cells acts to balance the water potential gradient across the cell membrane (*i.e.*, between the cytoplasm and ice formed within the apoplast).

In summary, negative turgor pressure remains an intriguing but little-studied area of plant water relations. While it does not appear to form in cells such as the leaf mesophyll, that can deform easily, its existence in living cells with either lignified cell walls or where the cell is embedded in a matrix of lignified tissues (*e.g.*, living cells within wood) cannot be ruled out. The potential benefits of negative turgor pressures in terms of preventing mechanical and osmotic damage associated with severe desiccation makes this topic worthy of further study.

THE MATRIC POTENTIAL

In discussions of soils, seeds, and cell walls, one often finds reference to yet another component of Ψ_w: the matric potential (Ψ_m). The matric potential is used to account for the reduction in free energy of water when it exists as a thin surface layer, one or two molecules thick, adsorbed onto the surface of relatively dry soil particles, cell walls, and other materials

. The matric potential does not represent a new force acting on water, because the effect of surface interactions can theoretically be accounted for by an effect on Ψ_s and Ψ_p. In dry materials, however, this surface interaction effect often cannot easily be separated into Ψ_p and Ψ_s components in dry materials, so they are frequently bulked together and designated as the matric potential.

It is generally not valid to add Ψ_m to independent measurements of Ψ_s and Ψ_p to arrive at a total water potential. This is particularly true for water inside hydrated cells and cell walls, where matric effects are either negligible or they are accounted for by a reduction in Ψ_p.

For instance, the negative pressure in water held by cell wall microcapillaries at the evaporative surfaces of leaves, discussed on page 45 of the textbook, is sometimes described as a wall matric potential. Care is needed to avoid inconsistencies when accounting for this physical effect in definitions of Ψ_p, Ψ_s, and Ψ_m.

MEASURING WATER POTENTIAL

Plant scientists have expended considerable effort in devising accurate and reliable methods for evaluating the water status of a plant. Four instruments that have been used extensively to measure Ψ_w, Ψ_s, and Ψ_p are described here: psychrometer, pressure chamber, cryoscopic osmometer, and pressure probe.

Psychrometer (Ψ_w Measurement)

Psychrometry (the prefix "psychro-" comes from the Greek word *psychein*, "to cool") is based on the fact that the vapour pressure of water is lowered as its water potential is reduced. Psychrometers measure the water vapour pressure of a solution or plant sample, on the basis of the principle that evaporation of water from a surface cools the surface.

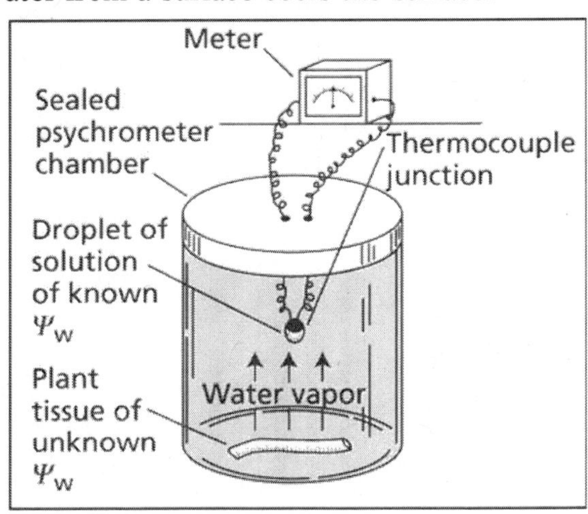

Fig. Diagram illustrating the use of Isopiestic Psychrometry to Measure the Water Potential of a Plant Tissue.

One psychrometric technique, known as *isopiestic psychrometry*, has been used extensively by John Boyer and coworkers. Investigators make a measurement by placing a piece of tissue sealed inside a small chamber that contains a temperature sensor (in this case, a thermocouple) in contact with a small droplet of a standard solution of known solute concentration (known Ψ_s and thus known Ψ_w). If the tissue has a lower water potential than that of the droplet, water evaporates from the droplet, diffuses through the air, and is absorbed by the tissue.

This slight evaporation of water cools the drop. The larger the difference in water potential between the tissue and the droplet, the higher the rate of water transfer and hence the cooler the droplet. If the standard solution has a lower water potential than that of the sample to be measured, water will diffuse from the tissue to the droplet, causing warming of the droplet.

Measuring the change in temperature of the droplet for several solutions of known Ψ_w makes it possible to calculate the water potential of a solution for which the net movement of water between the droplet and the tissue would be zero signifying that the droplet and the tissue have the same water potential.

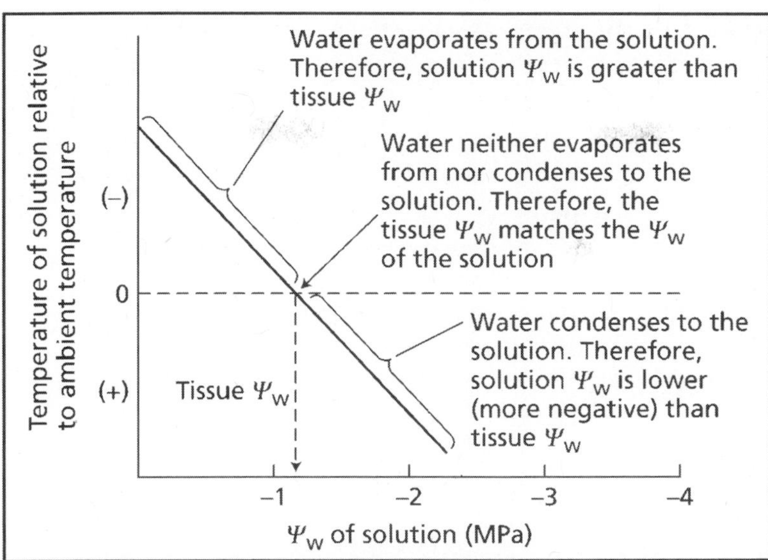

Psychrometers can be used to measure the water potentials of both excised and intact plant tissue. Moreover, the method can be used to measure the Ψ_s of solutions. This can be particularly useful with plant tissues. For example, the Ψ_w of a tissue is measured with a psychrometer, and then the tissue is crushed and the Ψ_s value of the expressed cell sap is measured with the same instrument. By combining the two measurements, researchers can estimate the turgor pressure that existed in the cells before the tissue was crushed ($\Psi_p = \Psi_w - \Psi_s$).

A major difficulty with this approach is the extreme sensitivity of the measurement to temperature fluctuations. For example, a change in temperature of 0.01°C corresponds to a change in water potential of about 0.1 MPa. Thus, psychrometers must be operated under constant temperature conditions. For this reason, the method is used primarily in laboratory settings.

Pressure Chamber (Ψ_w Measurement)

A relatively quick method for estimating the water potential of large pieces of tissues, such as leaves and small shoots, is by use of the pressure chamber. This method was pioneered by Henry Dixon at Trinity College, Dublin, at the beginning of the twentieth century, but it did not come into widespread use until P. Scholander and coworkers at the Scripps Institution of Oceanography improved the instrument design and showed its practical use. In this technique, the organ to be measured is excised from the plant and is partly sealed in a pressure chamber. Before excision, the water column in the xylem is under tension. When the water column is broken by excision of the organ (*i.e.*, its tension is relieved allowing its Ψ_p to rise to zero), water is pulled rapidly from the xylem into the surrounding living cells by osmosis. The cut surface consequently appears dull and dry.

To make a measurement, the investigator pressurizes the chamber with compressed gas until the distribution of water between the living cells and the xylem conduits is returned to its initial, pre-excision, state. This can be detected visually by observing when the water returns to the open ends of the xylem conduits that can be seen in the cut surface. The pressure needed to bring the water back to its initial distribution is called the *balance pressure* and is readily detected by the change in the appearance of the cut surface, which becomes wet and shiny when this pressure is attained.

The diagram at left shows a shoot sealed into a chamber, which may be pressurized with compressed gas. The diagrams at right show the state of the water columns within the xylem at three points in time: (A) The xylem is uncut and under a negative pressure, or tension. (B) The shoot is cut, causing the water to pull back into the tissue, away from the cut surface, in response to the tension in the xylem. (C) The chamber is pressurized, bringing the xylem sap back to the cut surface.

The pressure chamber is often described as a tool to measure the tension in the xylem.

However, this is only strictly true for measurements made on a non-transpiring leaf or shoot (for example, one that has been previously enclosed in a plastic bag). When there is no transpiration, the water potential of the leaf cells and the water potential in the xylem will come into equilibrium.

The balancing pressure measured on such a non-transpiring shoot is equal in magnitude but opposite in sign to the pressure in the xylem (Ψ_p). Because the water potential of our non-transpiring leaf is equal to the water potential of the xylem, one can calculate the water potential of the leaf by adding together Ψ_p and Ψ_s of the xylem, provided one collects a sample of xylem sap for determination of Ψ_s. Luckily Ψ_s of the xylem is usually small (> -0.1 MPa) compared to typical midday tensions in the xylem (Ψ_p of –1 to –2 MPa). Thus, correction for the Ψ_s of the xylem sap is frequently omitted.

Balancing pressure measurements of transpiring leaves are more difficult to interpret. The fact that water is flowing from the xylem to the leaf means that differences in water potential must exist. When the transpiring leaf or shoot is cut off, the tension in the xylem is instantly relieved and water is drawn into the leaf cells until the water potentials of the xylem and the leaf cells come into equilibrium.

Because the total volume of the leaf cells is much larger than the volume of sap in the xylem, this equilibrium water potential will be heavily weighted towards that of the leaf. Thus, any measurement of the balancing pressure on such a leaf or shoot will result in a value that is approximately the water potential of the leaf, rather than the tension of the xylem. (To be exact, one would have to add the Ψ_s of the xylem sap to the negative of the balancing pressure to get the leaf water potential.) One can explore the differences between the water potential of the xylem and the water potential of a transpiring leaf by comparing balancing pressures measured on covered (*i.e.*, non-transpiring) versus uncovered (transpiring) leaves.Pressure chamber measurements provide a quick and accurate way of measuring leaf water potential. Because the pressure chamber method does not require delicate instrumentation or temperature control, it has been used extensively under field conditions. For a more complete description of the theory and operation of the pressure chamber.

FLOWER STRUCTURE AND THE ANGIOSPERM LIFE CYCLE

The flower consists of several leaflike structures attached to a specialized region of the stem called the receptacle. Sepals and petals are the most leaflike. Petals have the primary function of attracting insects to serve as pollinators, accounting for their often showy and brightly coloured appearance. The stamen is the male sexual structure, and the pistil is the female sexual structure. The pistil is composed of one or more united carpels; the pistil, or in some flowers a whorl of pistils, is sometimes referred to as the gynoecium.

The stamen consists of a narrow stalk called the filament and a chambered structure called the anther. The anther contains tissue that gives rise to pollen grains. The pistil consists of the stigma (the tip where pollen lands during pollination), the style (an elongated structure), and the ovary. The ovary, the hollow basal portion of the pistil, completely encloses one or more ovules. Each ovule, in turn, contains an embryo sac, the structure that gives rise to the female gamete, the egg.

After landing on the stigma, the pollen grain germinates to form a long pollen tube, which penetrates the tissues of the style and ultimately enters the cavity of the ovary, which houses the ovule. Within the ovary, the pollen tube enters the ovule and deposits two haploid sperm cells in the embryo sac.

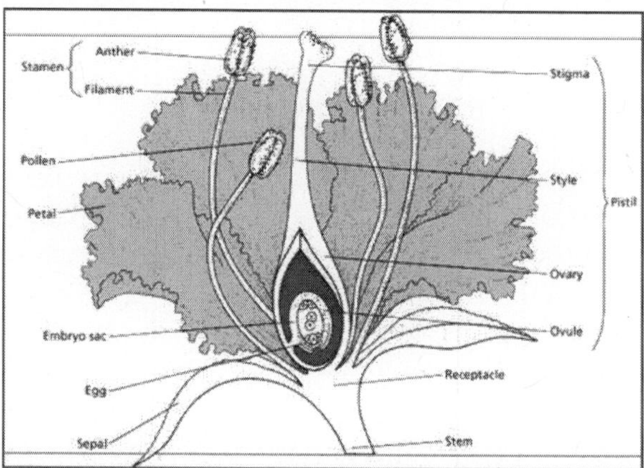

Fig. Schematic Representation of an Idealized flower of the Angiosperms.

One sperm cell fuses with the egg to produce the zygote; the other typically fuses with the two polar nuclei to produce a specialized storage tissue termed the endosperm, which provides nutrients to the growing embryo. Endosperm tissue also provides the bulk of the worldrquote s food supply in the form of cereal grains. As in conifers, in angiosperms the outer tissues of the ovule harden into a protective seed coat. Angiosperm seeds have a second layer of protective tissues, the fruit. The fruit consists of the ovary wall and, in some cases, receptacle tissue. Angiosperms are divided into two major groups, dicotyledons (dicots) and monocotyledons (monocots). This distinction is based primarily on the number of cotyledons, or seed leaves. In addition, the two groups differ with respect to other anatomical features, such as the arrangement of their vascular tissues, and their floral structure.

As the dominant plant group on Earth, and because of their great economic and agricultural importance, angiosperms have been studied much more intensively than other types of plants, and they are discussed extensively in this book. Plant physiologists have focused on a relatively small number of species that represent convenient experimental systems for the study of specific phenomena. Therefore, while we focus on these famous few, it is important to keep in mind the tremendous diversity of form and function that exists within the angiosperms, and the even greater diversity of form and function that is found within the plant kingdom as a whole.

PLANT TISSUE SYSTEMS: DERMAL, GROUND, AND VASCULAR

Dermal tissue. The epidermis is the dermal tissue of young plants undergoing primary growth. It is generally composed of specialized, flattened polygonal cells that occur on all plant surfaces. Shoot surfaces are usually coated with a waxy cuticle to prevent water loss and are often covered with hairs, or trichomes, which are epidermal cell extensions.

Pairs of specialized epidermal cells, the guard cells, are found surrounding microscopic pores in all leaves.

The guard cells and pores are called stomata (singular stoma), and they permit gas exchange (water loss, CO_2 uptake, and O_2 release or uptake) between the atmosphere and the interior of the leaf.

The root epidermis is adapted for absorption of water and minerals, and its outer wall surface typically does not have a waxy cuticle. Extensions from the root epidermal cells, the root hairs, increase the surface area over which absorption can take place.

Ground tissue. Making up the bulk of the plant are cells termed the ground tissue. There are three types of ground tissue: parenchyma, collenchyma, and sclerenchyma.

- Parenchyma, the most abundant ground tissue, consists of thin-walled, metabolically active cells that carry out a variety of functions in the plant including photosynthesis and storage
- Collenchyma tissue is composed of narrow, elongated cells with thick

primary walls. Collenchyma cells provide structural support to the growing plant body, particularly shoots, and their thickened walls are nonlignified so they can stretch as the organ elongates. Collenchyma cells are typically arranged in bundles or layers near the periphery of stems or leaf petioles.

- Sclerenchyma consists of two types of cells, sclereids and fibres. Both have thick secondary walls and are frequently dead at maturity. Sclereids occur in a variety of shapes, ranging from roughly spherical to branched, and are widely distributed throughout the plant. In contrast, fibres are narrow, elongated cells that are commonly associated with vascular tissues. The main function of sclerenchyma is to provide mechanical support particularly to parts of the plant that are no longer elongating.

In the stem, the pith and the cortex make up the ground tissue. The pith is located within the cylinder of vascular tissue, where it often exhibits a spongy texture because of the presence of large intercellular air spaces. If the growth of the pith fails to keep up with that of the surrounding tissues, the pith may degenerate, producing a hollow stem. In general, roots lack piths, although there are exceptions to this rule. In contrast, the cortex, which is located between the epidermis and the vascular cylinder, is present in both stems and roots.

At the boundary between the ground tissue and the vascular tissue in roots, and occasionally in stems, is a specialized layer of co rtex known as the endodermis. This single layer of cells originates from cortical tissue at the innermost layer of the root cortex and forms a cylinder that surrounds the central vascular tissue, or stele.

Early in root development, a narrow band composed of the waxy substance suberin is formed in the cell walls circumscribing each endodermal cell. These suberin deposits, called Casparian strips, form a barrier in the endodermal walls to the intercellular movement of water, ions, and other water-soluble solutes to the vascular cells. Leaves have two interior layers of ground tissue that are collectively known as the mesophyll. The palisade parenchyma consists of closely spaced, columnar cells located beneath the upper epidermis.

There is usually one layer of palisade parenchyma in the leaf. Palisade parenchyma cells are rich in chloroplasts and are a primary site of photosynthesis in the leaf. Below the palisade parenchyma are i rregularly shaped, widely spaced spongy mesophyll cells.

The spongy mesophyll cells are also photosynthetic, and the large spaces between these cells allow diffusion of carbon dioxide. The spongy mesophyll also contributes to leaf flexibility in the wind, and this flexibility facilitates the movement of gases within the leaf.

Vascular tissues: xylem and phloem. The vascular tissue is composed of two major conducting systems: the xylem and the phloem. The xylem transports water and mineral ions from the root to the rest of the plant. The phloem distributes the products of photosynthesis and a variety of other solutes throughout the plant. The tracheids and vessel elements are the conducting cells of the xylem. Both of these cell types have elaborate secondary-wall thickenings and lose their cytoplasm at maturity; that is, they are dead when functional. Tracheids overlap each other, whereas vessel elements have open end walls and are arranged end to end to form a larger unit called a vessel. Other cell types present in the xylem include parenchyma cells, which are important for the storage of energy-rich molecules and phenolic compounds, and sclerenchyma fibres.

The sieve elements and sieve cells are responsible for sugar translocation in the phloem. The former are found in angiosperms; the latter perform the same function in gymnosperms. Like vessel elements, sieve elements are often stacked in vertical rows, forming larger units called sieve tubes, whereas sieve cells form overlapping arrays. Both types of conducting cells are living when functional, but they lack nuclei and central vacuoles and have relatively few cytoplasmic organelles.

Substances are translocated from sieve cell to sieve cell laterally through circular or oval zones containing enlarged pores, called sieve areas. In contrast, sieve tubes translocate substances through large pores in the end walls of the sieve elements, called sieve plates. Sugar movement through sieve tubes is more efficient and rapid than through sieve cells and represents a more evolutionarily advanced mechanism. Sieve elements are associated with, and depend on, densely cytoplasmic parenchyma cells called companion cells. The analogous cells adjacent to the sieve cells of gymnosperms are called albuminous cells. Companion cells provide proteins and metabolites necessary for the functions of the sieve tube elements. In addition, the phloem frequently contains storage parenchyma and fibres that provide mechanical support.

ORIGIN AND BIOGENESIS OF MITOCHONDRIA

Earlier views about the origin of mitochondria from the plasma membranes about nuclear membranes are now discarded, for there is neither substance nor facts to substantiate such claims.

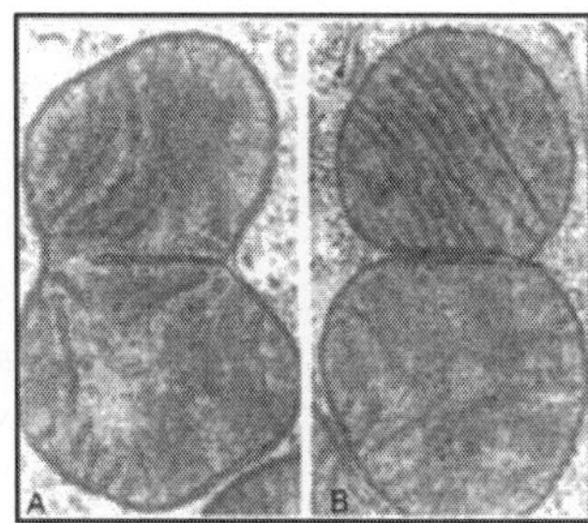

Fig. Mitochondrial Division

Recent studies have unequivocally showed that mitochondria take their origin in the cells from either pre-existing mitochondria or promitochondria. During mitochondrial biogenesis, genes of both nuclear genome and mitochondrial genome products interact. Though mitochondrial DNA codes for some 13 mitochondrial proteins, tRNA and rRNA, other components are coded for by nuclear genes, and they are synthesized in cytoplasm. Then they are transported into mitochondria through mitochondrial membrane transport complex, to form mitochondrial functional structures.

Vacuoles

Vacuoles are the spaces in the cells surrounded by single unit membranes. Don't consider them as empty spaces. In plants, mature cells have a single large centrally located vacuole occupies nearly 50-70% of the cellular volume. It is filled with cell sap. It is separated from the rest of the cytoplasm by a single unit membrane called tonoplast.

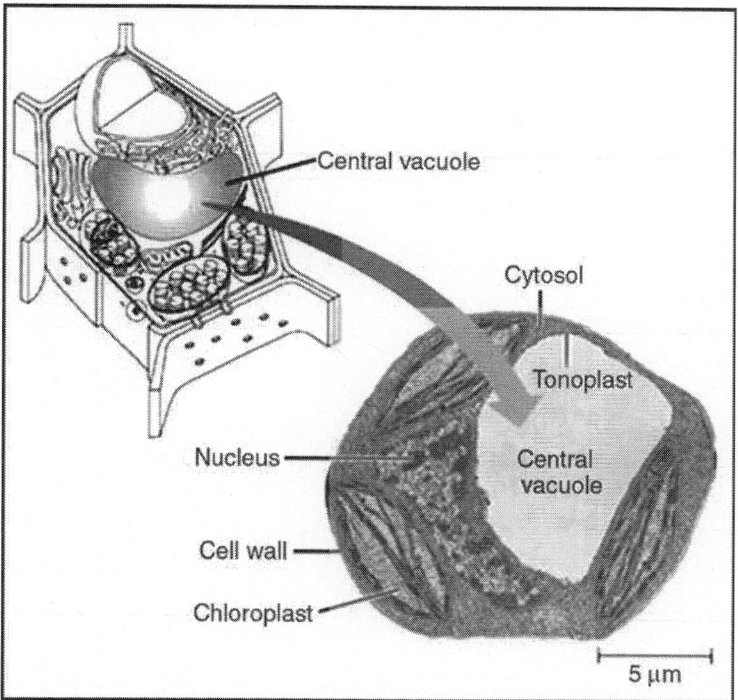

In meristematic cells, vacuoles are absent to being with, but as the cell expands, a number of small vacuoles arise from the endoplasmic reticulum; later as the cells further enlarge, smaller vacuoles fuse with one another to form a single large vacuole. The content of the cell vacuole is important in maintaining the turgidity of the cells. Furthermore, these vacuoles have been found to show lysosomal activities too. Nevertheless, vacuoles found in animal cells have drawn the attention, because they perform some important functions.

Contractile vacuoles and food vacuoles are notable among them. Contractile vacuoles are helpful in the excretion and others help in digesting the ingested food materials. Thus vacuoles play different roles in different organisms.

CYTOSKELETON

Microfilaments and Microtubules

Organisms, whether they are plants or animals, are made up of various types of cells, which have various shapes and functions. But all these cells are derived from the same mother cell called zygote. The shapes and functions of these cells is defined and determined at the time of differentiation (the topic is more complex).

Within these cells, cytoplasm with its components like Intermediate filaments (IF), microtubules (MT and microfilaments (MF) play important roles in determining cellular shape, structure and function. These structures act as the internal cytoskeleton and give a kind of mechanical strength to the cytoplasm and provided directions for the movement of cytoplasm, which makes the cytoplasm as a dynamic fluid.

Microtubules

Microtubules are fine tubular protein structures having dimensions of 200 A0 in diametre and many microns in length. The wall of the individual tubular structure is made up of 13 protein subunits called tubulins which exist in two forms called ∝and βsubunits; they are arranged alternatively.Hence the basic building blocks of microtubules are Tubulins of molecular weight 55000-58000 daltons. These subunits alternatively get polymerized at the membrane bound nucleating centres into tubular structures.

This polymerization can be inhibited by colchicine, an alkaloid extracted from the tubers of Colchicum autumnale.

Such microtubules are dispersed and distributed in different patterns in the cytoplasm. Some form a kind of network in the cytoplasm, some are oriented towards nuclear membrane, and some run parallel to the plasma membrane in longitudinal direction of the cell.

These structures being quite rigid, they are able to provide a structural support to various cells. Microtubule are associated with Endoplasmic reticulum for its sustainability and structural stability, so also golgi membranes.Not withstanding being the cytoskeleton structures, they are also involved in the structural organization of cilia, flagella, centrioles, basal granules, mitotic apparatus, neurotubules etc.

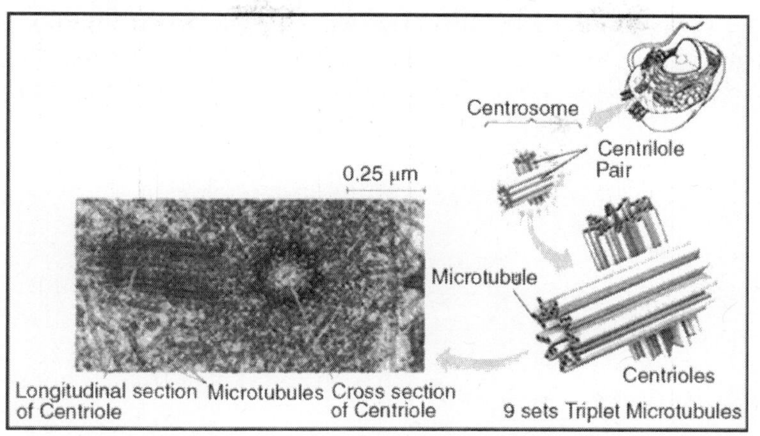

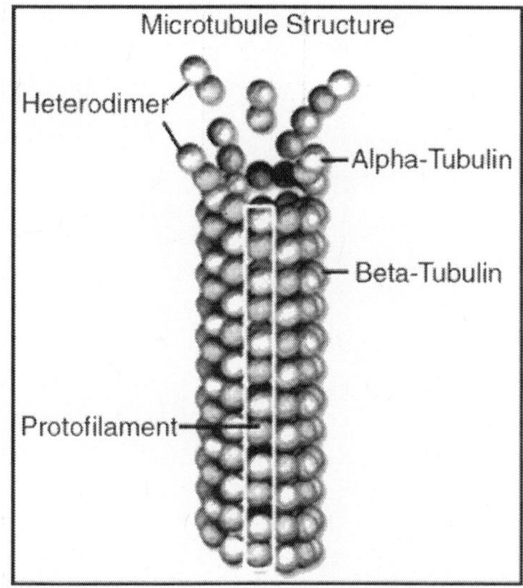

Such being the case, the function of not just being cytoskeleton support to the aqueous cytoplasm, but they are also involved in various functions like maintenance of the cell shape, cytoplasmic fluidity, membrane movement, chromosomal movement, cytokinesis, cell wall deposition, flagellar movement, sensory transduction, transport of cellular materials, cell polarity, morphogenesis and others to say the cytosolic goods to respective destinations and they aided by motor proteins such as kinesins, dyeinins and few more. They are involved in transport of goods along the MTs to their respective destinations. Tubulins are the most abundant proteins in the organic world next to RuBP carboxylase.

Microfilaments

Along with microtubules, cytoplasm also contains category of thin protein fibres called microfilaments. They represent the contractile system of the cytoplasm. Microfilaments are very fine protein filaments of 6-10 nm thickness and 10-100nm in length. These are made up of actin subunits and they can be associated with myosin and troponin proteins. Actin accounts for more than 20% of the total cellular proteins.

Actin consists of basic building blocks called globular proteins called G-actin, of molecular weight 45 Kd. These globular units, by undergoing polymerization, form into fibrous actin called F-actin. Normally two such F-actin proteins are helically coiled to each other and such filaments are associated with troponin.

These filaments are further decorated with: Myosin proteins with calcium binding sites and ATPase proteins.

These proteins are not just restricted to muscles, but they are also found in cytoplasm of almost all kinds of cells. The interaction between myosin and actin proteins generates the force for movement.

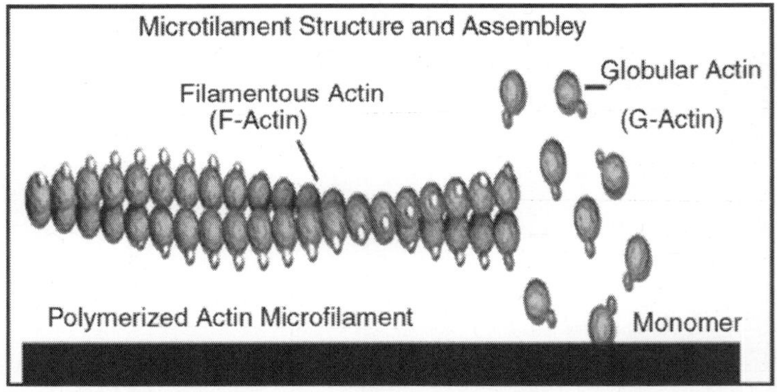

The transformation of G-Actin to F-Actin and vice versa represents the interconversion of sol-gel system of the cytoplasm is the dynamic force for cytoplasmic fluidity. The above said microfilaments are often located just beneath the plasma membranes in the form of bundles. These in turn are interconnected by a network of similar filaments which pervade the entire cytoplasm. These are also in contact with small membranous vesicles, microtubules, nucleus, polysomes, root lets of flagella etc. However mitochondria appear to be free from such filaments.

These microfilaments, particularly actin filaments undergo depolymerization in the presence of an alkaloid called cytochalasin B. If this drug is supplied to living cells, it inhibits many cellular activities like cytoplasmic streaming, migration of cells, endocytosis, exocytosis and even cell-polarity fixation. The above said features clearly suggest the involvement of microfilament is vital in various cellular activities. For example, protoplasmic streaming in many plasmodia, plant cells and amoeboid movement of many protozoans is controlled by the activity of these microfilaments. Recent investigations have shown that in cancer cells the microfilaments are in a disorganized state, which certainly indicates the involvement of microfilaments the induction of cancer.

Centrioles

Centrioles are the characteristic organelles of animal cells. Invariably they are present in all animal cells, but they are almost absent in plant cells with the exception of some lower unicellular algae. They are found in the cytoplasm at one pole of the nucleus. It is made up of a pair of dark granular structures called centrioles, which in turn are surrounded by an amorphous region called centrosphere. Each centriole is made up of two open barrel shaped structure measuring 0.2 um μ 0.5 um, sometimes they may be as long as 2um. These two structures are oriented at right angle to each other.

Structure

The wall of the open barrel shaped structure is made up of nine groups of microtubules arranged in a circle. Each group has three microtubules and skewed arrangement, towards the centre. These three microtubules are named A, B and C starting from the centre towards periphery. Compared to the proximal structure of the flagella, the central pair of tubules are absent.

However, in the case of centrioles, which develop into cilia, the wall is made up of 9 groups of tubules and each has two microtubules and there are two centrally located microtubules. Chemically centrioles are made up of proteins like tubulin, dynein (ATPase) and other contractile proteins. Added to this, the report about the presence of DNA is very interesting. However the chemical nature of the peripheral amorphous region is not known.

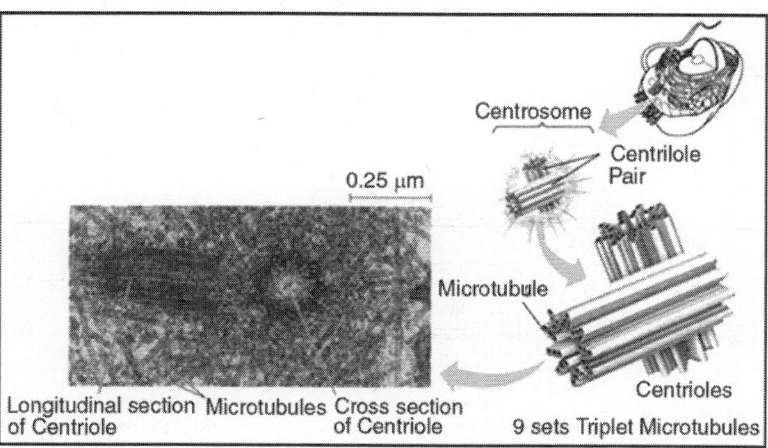

Centrioles have dual functions viz:

- Cilia and flagella in lower organisms and other cells take their origin from centrioles.

- During cell division, centrioles help in the organization of mitotic apparatus in a particular plane.

During cell division, particularly in animal cells, centrioles undergo duplication, so that centrioles produce procentrioles which are oriented at right angles to the mother centrioles.

The organization of microtubules during the formation of daughter centrioles is not clear. Whether the message for the synthesis of these proteins resides in the DNA found in centrioles or in the nucleus is not clear. How the peripheral cytoplasmic ribosomes are involved in the synthesis and secretion of these proteins is again a mystery. Nevertheless, during cell division the two centrioles separate, at the same time a number of microtubular structures emerge from each of the centrioles and radiate in all directions and inter link with the two centrioles. Thus, these centrioles move away from one another towards pre-determined polar directions. However, the mitotic fibres that develop during cell division do not take their origin from the centrioles.

2

Plant Cell Diagram and Functions

It is a commonly known fact that plants are highly eukaryotic organism. Their cells are membrane bound and are with many cell organelles. Even animals belong to eukaryoutic groups, but there are certain distinct differences between plant cell parts and animal cell parts. Plant cells possess very well developed cell walls, while an animal cell does not have that developed cell wall. Plant cell have many such distinct differences, which can be observed on studying the plant cell model. Cytoplasm and other organelles are very typical plant cell parts. Many scientific researches have been made to gain knowledge about various aspects of plants cell parts. Like every mechanical machine, which has dedicated parts for specific functions, plant cell parts also have highly specific and dedicated function.

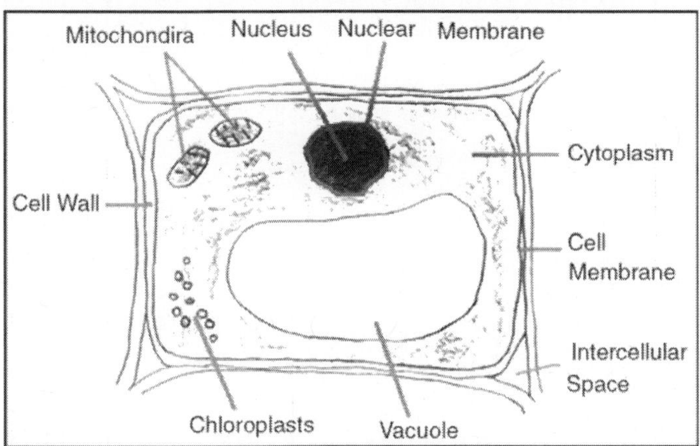

Fig. Labelled Plant Cell Diagram Sketch by Abhishake Sharma

A plant cell is actually a large microscopic 'fully functioning' city, literally and figuratively, housing miniature structures known as organelles.

It is fun to teach about a plant cell diagram and definitions, when your kid is already intrigued with the subject. Trouble is when he really isn't and would rather be doing anything else in the world than study biology. Yet, learning and teaching plant cell diagrams to children can actually be fun. You can use plant cell diagram for kids to make miniature models and the project will be fun as well as help your child learn.Kids love making and building things. So a plant cell diagram model set up on a sheet of thermoses, using interesting things to make up the internal organelles should really excite your child. Before you do that though, brush up your own knowledge of the labelled plant cell diagram and functions from this article. Know more on the plant cell structure and parts.

FUNCTIONS OF VARIOUS ORGANELLES

Let us begin with taking each internal organelle found in the plant cell diagram and highlighting its function and purpose within the plant cell. The most important thing to note are the differences that make a plant cell so different from an animal cell. A typical plant cell is distinguished from a typical animal cell as it contains things like the cell wall, vacuoles, chloroplasts and plasmodesmata, that the animal cells do not contain. Here's a look at a plant cell diagram with definitions of all the internal structures. Also know more on the similarities between plants and animal cells

This is a plant cell organelle, that store starch and is not found in all plant cells. It is usually only found in plants that are starchy in nature, like tubers and some fruits. These plastids are non-pigmented organelles that synthesize starch granules to convert into sugar, when the plant requires energy.

Alternately referred to as the ATP, this is a high energy molecule that stores up energy. ATP is produced by the plant cell in the cristae of the mitochondria and chloroplasts and supports the important function of energy transfer within the plant cells. This multifunctional nucleotide provides the energy for cellular processes like biosynthetic reactions and cell division.

Cell Membrane This is a thin wall that is found on the inside of the cell wall. It is a layer made up of protein, fats and cellular fibre, which provides the plant cell organelles with support and structure. Also known as the plasma membrane or plasmalemma, the cell membrane is the semi-permeable biological separation, between the cell insides and the cell outsides.

This is the think and rigid outer cover that lies above the cell membrane and surrounds the entire plant cell. This cell wall bonds with other plant cell walls and forms the structure of the plant that we know of. Made up of cellulose fibre, the cell wall is tough and acts as a filtering mechanism for the plant cell. Its most important function is to maintain internal plant cell pressure and prevent over-expansion when water enters the cell.

Also known as the microtubule organizing centre, the centrosome is 'centriole free' structure that has radiating tubules originating from a dense centre. Found near the nucleus of the plant cell, the centrosomes produce the microtubules that regulate the cell-cycle progression. When the plant cell divides (mitosis), the centrosome also divides into two parts that move in the opposite directions. If you draw a coloured plant cell diagram, you will have to make a green coloured molecular structure within, to show chlorophyll.

Chlorophyll is a molecule that carries on the process of photosynthesis. This is the process of producing sugar and oxygen using light energy, water and carbon dioxide. This usually green organelle is magnesium based and is found in many different molecular structures.

Chloroplast

This is usually the elongated or disc shaped photosynthesis site that contains the chlorophyll. Also a part of the 'plastids' group, chloroplasts are similar to mitochondria, but are only found in plants and protista. Chloroplasts have their own DNAs and are protected by the surrounding two lipid-bilayer membranes. Chromatin is a combination of DNA and protein that is highly complex in nature and is central to the makeup of the chromosomes. Found inside the nuclei of eukaryotic cells, the chromatin plays important roles to pack DNA into smaller volumes (to fit in the cell), control DNA replications and allow mitosis and meiosis.

Cristae are the finger-like projections that form the folded inner membranes of the mitocondria. ATP is generated in the walls of the cristae and they also help in cellular respiration. Made up of proteins, ATP synthase and various cytochromes, the cristae increase the surface area of the plant cell, on which the various reactions can actually take place. The cytoplasm is essentially the jelly-like substance outside the plant cell nucleus. This is where all the plant cell organelles are situated and it is the substance that is entirely encircled by the cell wall. The contents of the nucleus are called the nucleoplasm and are not part of the cytoplasm, but all the other plant cell organelles are are indeed a part of it. Know more about the structure and functions of cytoplasm.

Golgi Body

Also known as the golgi apparatus or the golgi complex, the golgi bodies are flat, layered sac-like organelles, that are located near the nucleus. These bodies do the important work of packing proteins and carbohydrates into membrane-bound vesicles. These vesicles are then exported from the cell.

Mitochondrian

A mitochondrian is a round or rod shaped organelle with a double layered membrane. The inner membrane is double folded and forms projections known as the cristae. The mitochondrian performs the essential task of converting energy stored in glucose form into ATP. Know more on mitochondria: structure and functions. This is spherical body that contains various organelles like the nucleolus (where ribosomal RNA is produced) and surrounded by a nuclear membrane. The nucleus is the control room that controls various cell functions by controlling the protein synthesis of the plant cell. The nucleus contains DNA within the chromosomes. Know more on cell nucleus: structure and functions.

These are the sites that see protein synthesis and are basically nothing but small cell organelles made up of RNA-rich cytoplasmic granules. These molecules make proteins out of amino acids. The word 'ribosome' is derived from the name 'ribonucleic acid' that is biologically very important for the plant cell.

Rough Endoplasmic Reticulum

Also known as rough ER, this is a vast interconnected system of membranes, infolded tubular structures and convulted sacs that are found in the cell's cytoplasm. This is a network that is covered with ribosomes and these are what give it a rough appearance. These networks transport materials through the cell and produce proteins in sacs known as cisternae. These are in turn sent to the golgi bodies and deposited into the cell membrane.

Smooth Endoplasmic Reticulum

Also known as smooth ER, this is the same thing as the rough ERs, but with the main difference that they are smooth. They bud off from the rough Ers and also transport materials within the cell. The ER channels, known as the ER lumen, are used to transport proteins and lipids to the golgi bodies and the membranes.

These are parts of the plant cell chloroplasts and are located between the grana within their inner cell membranes. A stroma is essentially the fluid matrix that surrounds the thylakoids.These are chloropyll containing disk shaped membrane structures found inside the chroloplasts. Chloroplasts are actually made up of a stack of thylakoid discs that aid in the process of photosynthesis. The stack of thylakoid disks in the chloroplasts are known as grana (singular granum) VacuoleThese are large fluid-filled, membrane-bound spaces within the plant cell.

They help in maintaining the cell shape and most plant cells have just one, single vacuole that represents upto 90 per cent of the total plant cell. They contain ions, sugar, secondary metabolites and enzymes. A vacuole is surrounded by a membrane called the tonoplast.here are many other small structures found within the plant cell like the intercellular air spaces (gap between two plant cells) and the peroxisome (has a chrystalline core and helps in removing hydrogen from substrates), but I will conclude my 'plant cell diagram' article here. Hope my overview of the plant cell diagram helps you with your task of educating your child with a plant cell model.

PLANT CELL ORGANELLES

Plants are highly evolved eukaryotic organisms that comprise of membrane bound cell organelles. Even though plants and animals belong to eukaryotic groups, they differ in certain characteristic features. For example, a plant cell possesses a well-developed cell wall and large vacuoles, while an animal cell lacks such structural parts. In addition to these, a plant cell lacks centrioles and intermediate filaments, which are present in an animal cell. The cells are the basic units of life that work together to perform life sustaining functions in both animal and plant worlds. Plant cells are eukaryotic cells having thick and rigid cell walls. Eukaryotic cells are cells that contain complex structures enclosed within membranes called nucleus or nucleus envelop, within which the genetic material is present. These cells are present in almost all living organisms including animals, plants and fungi.

PLANT CELL MODEL AND PARTS

Plant cells differ from animal cells in that they have three different structures known as cell wall, vacuoles and plastids. However, they lack centrioles and intermediate filaments which are present in animal cells.

Read more on plant cell structure and parts:

- *Cell Membrane*: It permits the waste material to exit the cell and regulates the movement of materials in and out of the cell. It also acts as a barrier between the inside of the cell and the outside, so that the chemical conditions on both sides can be different.
- *Cell Wal*l: It provides rigid structural support to the cell from which structures, like, leaves and stems are produced. It allows the circulation and distribution of water, minerals and nutrients in and outside the cell. The wall also controls the presence of pathogen microbes and development of tissues in the cell.
- *Golgi Body*: It stores, packages, and circulates the lipids and proteins made in the endoplasmic reticulum. It stores the proteins into packages called vesicles.
- Rough Endoplasmic Reticulum: It synthesizes and exports proteins and glycoproteins throughout the cell.
- *Lysosomes*: These organelles digest food by breaking down larger molecules into smaller ones, using special proteins.
- *Cytoplasm*: It helps in maintaining the cell shape. Along with this, it plays an important role in the internal movement of cell organelles, cell mobility and muscle fibre contraction. It also distributes oxygen and nutrients to different parts of the cell.
- *Nucleolus*: It is the most prominent structure in the nucleus wherein ribosomes are made.
- *Vacuole*: The plant cell contains a large, single vacuole (an enclosed compartment) which is used to store water, compounds and minerals that help in plant growth.
- *Ribosomes*: These are packets of RNA (Ribonucleic Acid) and are called as protein builders or synthesizers of the cell.

- *Chloroplasts*: They are the food producers of the cell, containing chlorophyll, the green pigment essential for photosynthesis. Their main function is to generate sugars and starches.
- *Nucleus*: It's the most important part of the cell, comprising chromosomes, *i.e.* structures made up of genetic information that helps in cell growth and reproduction. Read more on cell nucleus: structure and functions.
- *Nuclear Envelope*: It's an enclosure that surrounds nucleus and its contents. Unlike cell membrane which has pores and spaces for RNA and proteins to pass through, it keeps the chromatin and nucleolus inside the nucleus.
- *Smooth Endoplasmic Reticulum*: It's main function is to package proteins for transport, synthesize membrane phosolipids, and secrete calcium. It also performs transformation of bile pigments, glycogenolysis (the breakdown of glycogen), and detoxification of different drugs and chemical agents.
- *Mitochondria*: It provides energy to the cell by combining sugar molecules with oxygen to generate carbon dioxide and water.
- *Amylosplast*: It's an organelle present in some plant cells that stores starch.
- *Druse Crystal*: It's a granular type of crystal found in plant vacuoles. It is composed of calcium oxalate and is considered to deter herbivory.
- *Centrosome*: Also known as Microtubule Organizing Centre, it's a region in the cell where microtubules are produced that perform a variety of functions ranging from transportation to structural support. Though both plant and animal cell centrosomes play similar roles during cell division, plant cell centrosomes are simpler and do not contain centrioles.
- *Peroxisomes*: These are membrane bound packets of oxidative enzymes that convert fatty acids into sugar and assist chloroplasts in photo-respiration.

A typical plant cell is made up of cytoplasm and organelles. Scientific studies have been done regarding plant cell organelles and their functions. Each of the organelles of a plant cell has specific functions, without which the cell cannot operate properly.

Cytoplasm is the thick, gel-like semitransparent fluid that is found in both plant and animal cell. It is bounded by the plasma membrane, and contains many organelles in an eukaryotic cell (cell containing membrane bounded nucleus). In the eukaryotic cell, the nucleus is separated from the cytoplasm by a double membrane, known as nuclear membrane. Cytoplasm was discovered in 1835. Though, discovery of the different organelles found in cytoplasm can be attributed to different scientists, no single scientist can be credited for discovering cytoplasm.

STRUCTURE OF CYTOPLASM

Basically, cytoplasm is the fluid, where the organelles remain suspended. So, it fills up the cell, especially the spaces not occupied by any organelle. The constituent parts of cytoplasm are cytosol, organelles and cytoplasmic inclusions.

Cytosol

Cytosol is the part of the cytoplasm that is not occupied by any organelle. It is a gelatinous substance. It mainly consists of cytoskeleton filaments, organic molecules, salt and water. Cytoskeleton filaments are made up of protein filaments, and they are responsible for giving the shape to the cell. It also contains enzymes, fatty acid, sugar and amino acid. Besides, ribosomes, proteasomes and soluble proteins can also be found in cytoplasm. Cytosol accounts for almost 70% of the total cell volume.

Organelles

Cytoplasm also contains some important organelles like mitochondria, endoplasmic reticulum, lysomes and Golgi apparatus. Besides, it also contains chloroplast in plant cells. Each organelle is bounded by a fatty membrane, and has some specific functions.

Cytoplasmic Inclusions

Some insoluble suspended substances found in cytosol are known as cytoplasmic substances. They are basically granules of starch and glycogen, and they can store energy. Besides, crystals of some minerals and lipid droplets can also be found in cytoplasm. Lipid droplets are composed of lipids and proteins, and they act as a storage of fatty acid and steroids.

Functions of Cytoplasm

Cytoplasm is the site of many vital biochemical reactions crucial for maintaining life. It is the place where cell expansion and growth take place. It provides a medium in which the organelles can remain suspended. Besides, cytoskeleton found in cytoplasm gives the shape to the cell, and facilitates its movement. It also assists the movement of different elements found within the cell. The enzymes found in the cytoplasm breaks down the macromolecules into small parts so that it can be easily used by the other organelles like mitochondria. For example, mitochondria cannot use glucose present in the cell, unless it is broken down by the enzymes into pyruvate. They act as catalysts in glycolysis, as well as in the synthesis of fatty acid, sugar and amino acid. Cell reproduction, protein synthesis, anaerobic glycosis, cytokinesis are some other vital functions that are carried out in cytoplasm. Organelles present in cytoplasm have some specific vital functions; for example, the function of mitochondria is to produce and store energy, while endoplasmic reticulum facilitates synthesis and transport of protein, production of steroids, production as well as storage of glycogen, etc.

The important functions of Golgi apparatus include modification, packaging, transportation and processing of macromolecules, like proteins, and lipids. On the other hand, lysosomes contain digestive enzymes and hence, they digest food particles, damaged or worn-out organelles and also virus and bacteria. However, the smooth operation of all these functions depend on the existence of cytoplasm, as it provides the medium for carrying out these vital processes.

LIST OF PLANT CELL ORGANELLES

When it comes to plant cell organelles, they are more or less similar to animal cells, except that the latter lacks chloroplast organelles, that are responsible for photosynthesis. Following is a list of organelles found in plant cell:

Nucleus

Nucleus (plural nuclei) is a highly specialized cell organelle, which stores the genetic component (chromosomes) of the particular cell. It serves as the main administrative centre of the cell, by coordinating the metabolic processes like cell growth, cell division and protein synthesis. Read more on cell nucleus.]

Plastids

Plastids are collective terms for organelles that carry pigments. In a plant cell, chloroplasts are the most prominent forms of plastids, that contain the green chlorophyll pigment. Because of these chloroplast plastids, a plant cell has the ability to undergo photosynthesis in the presence of sunlight and synthesize its own food. Read more on the importance of photosynthesis.

Ribosomes

Ribosomes are plant cell organelles that comprise of proteins (40 per cent) and ribonucleic acid or RNA (60 per cent). They are important organelles responsible for the synthesis of proteins. Each ribosome consists of two parts, a larger subunit and a smaller subunit.

Mitochondria

Mitochondria (singular mitochondrion) are spherical to rod-shaped organelles present in the cytoplasm of the plant cell.

They break down the complex carbohydrates and sugars into usable forms, for the plant. As mitochondria indirectly supply energy for the plant cell, they are also called as the powerhouse of the cell.

Golgi Body

Golgi body is also referred to as golgi complex or golgi apparatus. It plays a major role in transporting chemical substances in and out of the cell. After the endoplasmic reticulum synthesizes lipids and proteins, golgi body alters and prepares them for exporting outside the cell.

Endoplasmic Reticulum

Endoplasmic reticulum (ER) is the connecting link between the nucleus and cytoplasm of the plant cell. Basically, it is a network of interconnected and convoluted sacs that are located in the cytoplasm.

Based on the presence or absence of ribosomes, ER can be of smooth or rough types. The former type lacks ribosomes, while the latter is covered with ribosomes. Overall, endoplasmic reticulum serves as a manufacturing, storing and transporting structure for glycogen, proteins, steroids and other compounds.

Vacuoles

Vacuoles are the storage organelles that help in regulating turgor pressure of the plant cell. In a plant cell, there can be more than one vacuole. However, the centrally located vacuole is larger than others, which stores all sorts of chemical compounds. Vacuoles also assist in intracellular digestion of complex molecules and excretion of waste products.

PLANT GROWTH IN DIFFERENT WAYS

Growth in plants is defined as an irreversible increase in volume. The largest component of plant growth is cell expansion driven by turgor pressure. During this process, cells increase in volume manyfold and become highly vacuolate. However, size is only one criterion that may be used to measure growth. Growth also can be measured in terms of change in fresh weight—that is, the weight of the living tissue—over a particular period of time. However, the fresh weight of plants growing in soil fluctuates in response to changes in water status, so this criterion may be a poor indicator of actual growth. In these situations, measurements of dry weight are often more appropriate. Cell number is a common and convenient parameter by which to measure the growth of unicellular organisms, such as the green alga *Chlamydomonas*. In multicellular plants, however, cell number can be a misleading growth measurement because cells can divide without increasing in volume.

Growth is assessed by a count of the number of cells per milliliter at increasing times after the cells are placed in fresh growth medium. Temperature, light, and nutrients provided are optimal for growth. An initial lag period during which cells may synthesize enzymes required for rapid growth is followed by a period in which cell number increases exponentially.

This period of rapid growth is followed by a period of slowing growth in which the cell number increases linearly. Then comes the stationary phase, in which the cell number remains constant or even declines as nutrients are exhausted from the medium.

For example, during the early stages of embryogenesis, the zygote subdivides into progressively smaller cells with no net increase in the size of the embryo.

Only after it reaches the eight-cell stage does the increase in volume begin to mirror the increase in cell number. Because the zygote is an especially large cell, this lack of correspondence between an increase in cell number and growth may be unusual, but it points out the potential problem in equating an increase in cell number with growth.

Although cell number may not always be a reliable measure of plant growth, under most circumstances dividing cells, particularly in meristems, double in volume during their cell cycle. Therefore, an increase in cell number, such as the increase brought about by the activity of the apical meristems, does contribute to plant growth. However, the largest component of plant growth is the rapid cell expansion that occurs in the subapical region after cell division ceases.

Because all the cells of the plant axis elongate under normal conditions, the greater the number of cells produced by the apical meristem, the longer the axis will be. For example, when Arabidopsis plants are transformed with a gene that encodes cyclin, a key component of the cell cycle regulatory machinery, the cells of the apical meristem progress through their cell cycles more rapidly, so more cells form per unit time.

As a result, the roots of these transgenic plants have more cells and are substantially longer than the roots of wild-type plants grown under similar conditions. New cells form continually in the apical meristems. With each new round of cell division and associated cell expansion, the older derivatives are displaced a small distance from the apex. As the cells recede farther from the apex, the rate of displacement is greatly accelerated. By viewing plant growth as a process of cell displacement from the apex, we can apply the principles of kinematics.

3

Primary Growth of Plant

The primary growth of a plant body is brought about by the activity of primary meristems, particularly by apical meristem.

Details about primary growth occurring in stems, leaves and roots, may be studied by observing a transverse structure is differentiated into outer epidermis, the inner ground tissue and vascular tissue.

- *Dicotyledonous stem*: Example: Sunflower (Helianthus annus). The transverse section is almost circular in outline.
- *Epidermis*: This is the outermost layer, uniseriate, covered with thin cuticle, and bears a few stomata and many multicellular trichomes.
- *Ground Tissue*: This fundamental tissue is differentiated into outer cortex and central pith.
- *Cortex*: This is many layered and differentiated into hypodermis and inner cortex. Hypodermis is found below the epidermis, and is constituted with angular collenchyma which is 3-5 layered.

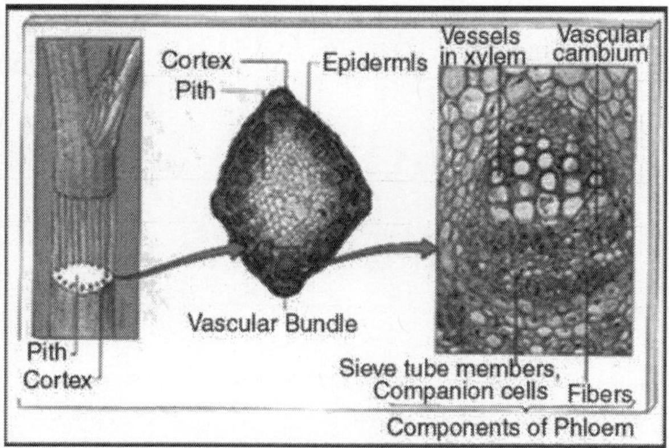

The inner part of cortex is constituted with chlorenchyma and parenchyma; large numbers of intercellular spaces are present between these cells. The inner most cortical layer is wavy in outline, rich in starch grains, with barrel shaped cells and is called endodermoid layer. (Endodermis is generally absent in stem; endodermoid layer differs from endodermis in lacking casparian thickenings).

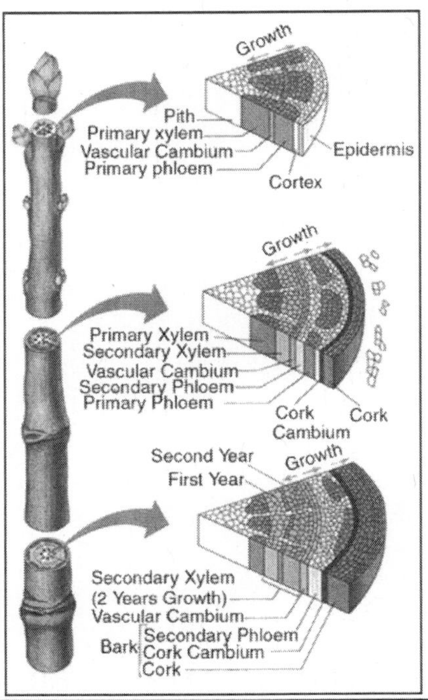

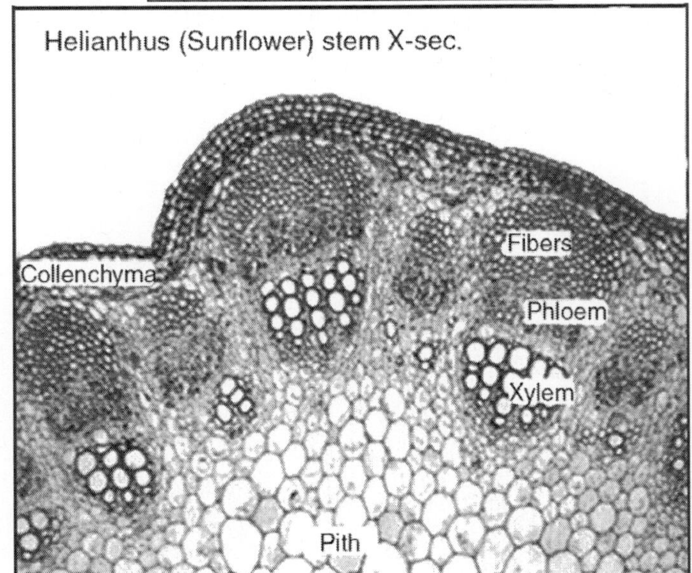

- *Pericycle*: This is the outermost region of vascular tissue. This consists of alternately arranged patches of sclerenchyma and parenchyma. The sclerenchyma is also known as hard bast fibres. Since it is found external to vascular bundle, the patch of sclerenchyma is also called bundle cap.

- *Vascular tissue*: This consists of many vascular bundles arranged regularly in the form of broken ring (Eustele). In each vascular bundle the xylem and phloem elements are in the same line; the presence of one to few stripped intrafascicular cambium is found in between xylem and phloem; the cambial cells are thin walled and are almost rectangular phloem is only external to xylem: the protoxylem is nearest to centre. The vascular bundles are conjoint, collateral, endarch, open and wedge shaped.
- *Pith or Medulla*: This is inner to vascular tissue and occupies the central region. It is composed of cells of parenchyma, with intercellular spaces. Extensions of these cells into the areas in between vascular bundles are called medullary rays.
- *Monocotyledonous stem*:
  Example: Zea mays (Maize). The outline is circular.

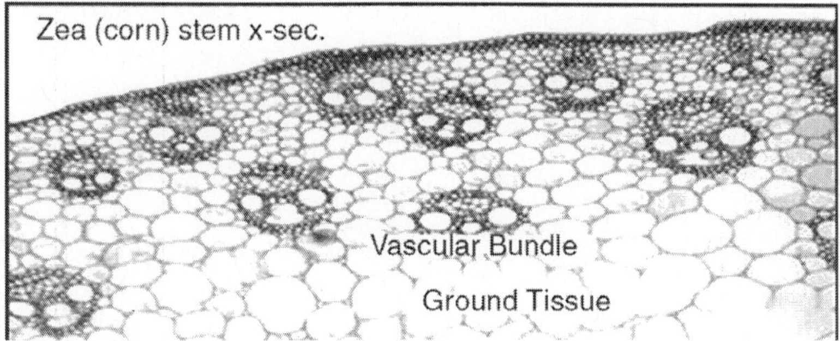

- *Epidermis*: This is the uniseriate, outermost layer and is covered with thick cuticle. Trichomes are absent.
- *Ground tissue*: The ground tissue is not differentiated; into cortex and pith, since there is scattered arrangement of vascular bundles. Even endodermoid layer and pericycle are also not differentiated. Hypodermis constitutes the exterior ground tissue, and is composed of 2 – 3 layers of sclerenchyma. The major part of the ground tissue is parenchymatous.
- *Vascular tissues*: This consists of numerous, oval shaped vascular bundles scattered throughout the ground tissue, hence the arrangement of vascular tissue is called atactostele. Peripheral vascular bundles are small-sized but many in numbers. The central vascular bundles are large sized but a few in numbers in all the vascular bundles the xylem and phloem are in the same line; cambium is absent; phloem is only external to xylem; protoxylem faces the centre.

Hence the vascular bundles are described as conjoint, collateral, endarch and closed. Each vascular bundle is surrounded by a bundle sheath of sclerenchyma, which is prominent at the base and tip of vascular bundle. Xylem comprises usually 2 oppositely arranged Metaxylem elements (pitted, facing periphery) and a few protoxylem elements (facing the centre of stem) in a linear row.If an imaginary line is drawn connecting the two Metaxylem elements with that of protoxylem, xylem appears 'Y' shaped.Usually the lower protoxylem elements break down and form a lysogenous cavity. Above the metaphloem, a small band of obliterated protophloem is seen. Metaphloem consists of sieve tubes and companion cells but not phloem parenchyma.

Comparison between the stems of Dicot and Monocot:

		Dicot stem *Example*: **Sunflower**	**Monocot stem** *Example*: **Maize**
1.	Epidermal trichomes	Present multicellular	Absent
2.	Hypodermis	Collenchymatous	Sclerenchymatous
3.	Ground tissue	Differentiated into cortex and Pith	Undifferentiated
4.	Endodermoid layer and Peri-cycle	Present	Absent
5.	Stele	Eustele	Atactostele
6.	Vascular bundles	Many open, wedge- shaped	Numerous, closed, oval-shaped
7.	Lysigenous cavity	Absent	Present
8.	Bundle cap	Present	Absent
9.	Bundle sheath	Absent	Present
10.	Secondary Growth	Occurs later	Does not occur

Dicotyledonous root: Example: *Lablab purpureus (Beans)*:

- *Epiblema*: The external protective layer is called Epiblema or Piliferous layer (the term epidermis is generally not applied to roots). It is uniseriate. Cells are thin walled. Some of these cells extend into unicellular root hairs.
- *Cortex*: The region inner to epiblema is called cortex and it is homogenous. The cortex is composed of thin walled, spaciously arranged parenchyma cells that constitute many layers.

  With increase in age, the exterior cortical region becomes suberized and dead. This exterior cortex is known as exodermis.

The innermost cortical layer is called endodermis. Its cells are barrel-shaped and are closely packed. Many cells get suberized on their radial and inner walls. These thickenings are called casparian stripes. Such endodermal cells may become totally suberized on all wall surfaces and are called stone cells.

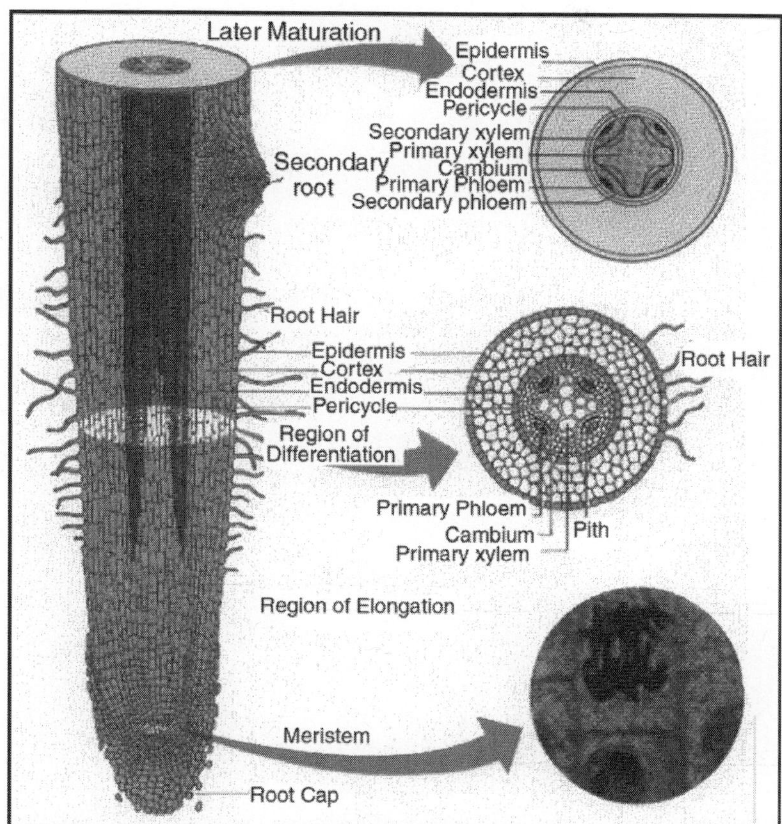

These cells do not allow conduction. The non-suberized endodermal cells are called passage cells. Inner to endodermis, one or two layers of parenchyma are present, constituting pericycle.

- *Stele*: Xylem and phloem patches are in equal number and alternate with each other. Xylem or phloem patches are usually four in number. In Xylem, the protoxylem elements face pericycle; between xylem and phloem elements, parenchyma cells are present constituting conjunctive tissue, which later becomes a secondary meristem.

  Hence vascular bundles are described as radial, exarch and open. Pith is very small, with closely packed parenchyma cells.

  Monocotyledonous root:

  Example: Canna

  Outline almost circular.

The nature of epiblema, exodermis, cortex, endodermis and pericycle is similar to that of Dicot root. However in cortex, large air spaces of schizogenous origin may be seen. The nature and arrangement of vascular bundles are also similar to that of Dicot root. But vascular bundles are usually more than six in number. The Parenchyma present in between xylem and phlegm is called conjunctive tissue, with thin cell walls. This tissue does not become meristematic and thus differs from Dicot root. Pith is relatively large and parenchymatous.

Pith and conjunctive tissue, become lignified and Sclerenchymatous. The vascular bundles are described as polyarch, radial, exarch and closed. An endogenous lateral branch may be seen developing from pericycle and passing through cortex.The leaf is horizontal and dorsiventrally differentiated (upper = ventral, lower = dorsal).

- *Epidermis*: The two surfaces are bounded by their respective epidermal layers. The epidermal cells are barrel-shaped, compactly arranged; upper epidermis is covered with thick cuticle and lacks stomata; lower epidermis is light green, covered with thin cuticle and is interrupted by stomata.
- *Mesophyll*: This is the ground tissue. It is differentiated into upper palisade and lower spongy parenchyma. Palisade parenchyma is found immediately below the upper epidermis, 2 to 3 layered, with compactly arranged tubular cells, rich in parietal chloroplasts.
- Spongy parenchyma is found above the lower epidermis; these cells are varied in shape and sizes, very loosely arranged enclosing air spaces some of which open into stomata. Chloroplasts are parietal in the parenchyma cells.

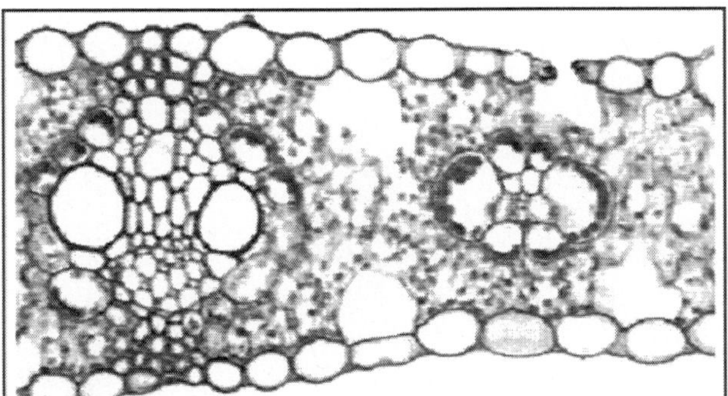

- *Vascular tissue*: Vascular bundles vary in size; each bundle is conjoint, collateral and closed. The vascular bundle is covered by a bundle sheath of parenchyma cells. An extension of bundle sheath is seen above and below the vascular bundle, reaching the epidermal layers. Phloem is towards lower epidermis; xylem is towards upper epidermis, with metaxylem facing phloem. Fibres are absent in both xylem and phloem. The xylem and phloem elements are conspicuous only in large vascular bundles.

Monocotyledonous leaf Example: Maize.

The leaf is Isobilateral in nature and is vertically oriented.

- *Epidermis*: This is uniseriate, with barrel-shaped, compactly arranged cells and is covered with thick cuticle.

  Stomata are found on both upper and lower epidermal layers hence it is called amphistomatic (more on the lower epidermis). Though the leaf is referred to as isobilateral, it is only in upper epidermis, a few large, empty and colourless bulliform or motor cells are present. During dry weather, these motor cells help the leaf to role over, due to the changes in turgidity. This rolling of leaf reduces the rate of stomatal transpiration.
- *Mesophyl*l: There is no differentiation of mesophyll into spongy and palisade parenchyma. All the cells of chlorenchyma are alike, isodiametric, almost compactly arranged with numerous parietal chloroplasts.

Table. Comparison Between the Leaf Structures of Dicot and Monocot.

	Dicot Root Ex: Mango	**Monocot Root Ex : Maize**
1. Orientation of leaf	Horizontal, dorsiventrally differentiated	Vertical, isobilateral
2. Stomata	Usually on lower epidermis	On both epidermal layers
3. Cuticle	Thick on upper epidermis	Thick on both epidermal layers
4. Motor cells	Absent	Present in upper epidermis
5. Mesophyll	Differentiated into upper palisade parenchyma and lower spongy parenchyma	No differentiation of mesophyll
6. Vascular bundles	All of them are not seen in parallel series and are supported by bundle sheath extension	Vascular bundles are in parallel series and are supported by sclerenchyma patches

- *Vascular tissue*: The vascular bundles are numerous, arranged in parallel series (venation is palmate-parallel), conjoint, collateral and closed. Phloem is towards lower epidermis. Each vascular bundle is surrounded by chlorenchymatous bundle; this sheath also serves for temporary storage of starch. A few vascular bundles are larger in size, with more amount of xylem and phloem and with large bundle sheath cells. A patch of sclerenchyma is present above and below the large sized vascular bundles.

PLANT GROWTH INTERACTIONS

Soil is a medium for plant growth, and its structure should not hinder the movement of water, oxygen and nutrients to plant roots, impede the growth of roots, or allow the buildup of toxic substances around roots. Soil structure can be described as the organization of the particles in the soil—the internal configuration of the soil matrix. Soil structure is formed by many interacting processes in the soil, especially shrinkage with drying, channel formation by plant roots, faunal activity, and cultivation.

Well-developed soil structure is represented by distinct stable aggregations (peds) of soil particles separated by interped pores. The interped pores of an aggregated soil are larger, on average, than pores between particles of unaggregated soil, resulting in a less dense condition.

The larger pores are also critical in terms of allowing movement of air and water through the soil. Unfavourable soil structure, on the other hand, can impart the characteristics of greater soil density, decreased aeration, poor percolation, and mechanical impedance to roots, which can restrict plant growth. There are a number of factors that may result in poor soil structure. Soil texture is important; in fact, very sandy soils are usually classified as structureless due to lack of soil matrix (sand is viewed more as the "skeleton" of the soil). In sandy soils, size distribution and packing of the grains are considered rather than structure per se. In loamy and finer-textured soil, clay acts as a cementing agent to provide some level of stability to the particle orientation in structured soil; however, clay can also cement particles in a structurally degraded condition ("massive"). Chemical conditions can be responsible for structurally degraded soil; a relatively high sodium content causes dispersion of particles and elimination of aggregation. These "sodic soils" occur naturally but can also be initiated or exacerbated by human activity. The human-imposed factor that degrades soil structure to the greatest extent is compaction. This chapter focuses on the effects of compacted soil on plant growth, including the direct effect of mechanical impedance as well as the indirect effects that usually accompany compaction. Finally, amelioration of compaction is discussed, with emphasis on the ability of plant roots to effect changes in soil structure and hence modify physical properties.

COMPACTION

Hillel described soil compaction as the compression and densification of an unsaturated soil body, reducing the fractional air volume. The degree to which a soil will compress is a function of the soil moisture content, antecedent bulk density, magnitude of the compactive effort, soil texture, and content of organic matter. Forces that compact soil can originate from natural sources, including rainfall, radial root pressure, cycles of wetting/drying and freezing/thawing, or traffic of people and machinery. The potential to degrade soil structure has increased as modern machinery has become larger and heavier, causing compaction to a greater depth. Compounding this problem is the tendency to tread repeatedly over soil while implementing cultural practices such as seeding, fertilization, pest control, harvesting/mowing, or construction during site development.

Soil compaction on recreational sites is thought to be one of the most challenging issues faced by turfgrass managers. Cultivation implements are generally designed to overcome the bonds stabilizing structure, breaking aggregates into smaller ones for seedbed preparation to maximize seed-soil contact, enhance water infiltration and retention characteristics, and improve soil aeration. Routine tillage of soil, however, can also degrade soil structure by shattering aggregates, eliminating pore networks, and imposing compacting forces, particularly in a subsurface layer of soil.

Soil Responses

Porosity, bulk density, and water retention characteristics are physical properties characterizing soil structure and compaction phenomena. Compaction of a soil decreases soil porosity and subsequently increases soil bulk density. Veihmeyer and Hendrickson showed that the critical soil bulk density needed to inhibit sunflower (Helianthus annus L.) root growth varied with texture; no roots penetrated soil with a bulk density of 1.9 g cm^{-3}. The lowest density where roots failed to penetrate was 1.46 g cm"3 for an Aiken clay loam. Air tends to occupy the larger pores of soil, whereas the affinity of water for mineral surfaces creates water films and wedges (menisci) within the small fissures and capillaries of the soil.

A reduction in the total porosity of soil, as a result of compaction, generally occurs at the expense of the air porosity, those relatively large pores not filled with water at a given soil water potential (usually "33 kPa to represent "field capacity").

Swartz and Kardos observed a decrease in air porosity from 19.3 to 12.7%, with increasing levels of compaction on several sand-soil-peat mixes differing in moisture content at the time of compaction, whereas capillary porosity increased from 32.2% at the lowest compaction level to 35.7% at the highest level.

In aggregated soils, the loss of porosity due to compaction can be attributed to the collapse of interaggregate pores. Water retention and flow characteristics in soil are influenced by alterations of soil porosity. Relative to Non-compacted soil, compacted soil will generally retain a greater amount of water at a given soil matric potential as a result of increased capillary porosity. Soil water permeability, particularly at saturated or nearly saturated conditions, is a function of air porosity; as air porosity decreases, so does water permeability. Compacted soils with lower infiltration rates are prone to problems with stagnant water and anaerobic conditions.

Constraints on Root Growth

Soil Aeration

Soil oxygen movement to plant roots is critical to maintain adequate respiration for growth. Anaerobiosis, or oxygen stress, occurs in soil when the rate of supply falls below the biological demand. Impeded soil aeration can arise from poor drainage and waterlogging or from compacted soil. One approach to evaluate soil aeration is to measure the composition of soil air. This technique indicates that gas exchange between the atmosphere and soil air is restricted when oxygen content of the soil air falls significantly below that of the atmosphere. An important challenge with this approach is how to extract a representative sample of soil air that avoids mixing of gases from outside the point of sampling or contamination from the atmosphere.

Another approach for assessing soil aeration is to determine the air porosity, or fractional air space, at a standard soil water potential. Work to identify threshold air porosity values below which plant growth is limited has generated values ranging from 5 to 20%. Madison has indicated that 10% air porosity is the threshold level for intensively utilized turf. The United States Golf Association Green Section (USGA) has suggested 15% air porosity at a 3 kPa water tension as the lower test limit for materials used in the construction of root zones for golf course putting greens. The extent to which air porosity in the root zone changes with time under a perennial turf system is a subject of current studies.

The rate of air exchange, rather than the volume of soil air, is the more important factor in soil aeration and therefore limits the usefulness of air porosity as an index of soil aeration. Gaseous movement in the soil occurs by two main processes: convection and diffusion. Convection, where the moving force is a gradient of total gas pressure, is thought to be a minor mechanism of soil aeration except at shallow depths and in soils with large pores. Although convection is not considered an important factor for more deeply rooted plant species, it could be more influential for shallow-rooted turfs. A large fraction of the root system of moderately to highly maintained turfs is found at surface depths of 5 cm or less.

Furthermore, the crown, the major meristematic organ, of turfgrass plants is located near the soil surface. Thus, turfs with a high plant population could have sufficiently high oxygen demands at the soil surface that convection would play a useful role in maintaining adequate aeration for plant crowns and surface rooting. Convection, therefore, could be partially responsible for the observations of improved turfgrass quality after changes in air masses (pressure) associated with passing weather systems. Below the surface, however, diffusion is considered the more important exchange mechanism in soil, and the moving force is a gradient of partial pressure. Gas exchange between the atmosphere and soil is maintained predominantly by diffusion through air-filled pores. In contrast, live tissue is typically hydrated, and the supply of oxygen to roots occurs by diffusion through water films. This is an important distinction, since the diffusivity of oxygen in water is less than its value in air by a factor of 10"4.

The solubility of oxygen in water may be a compounding issue, especially at higher temperatures, where solubility decreases. Methods that measure gaseous diffusion through air-filled pores do not provide information on the impedance to oxygen presented by the water film surrounding a root. It is possible that an inadequate supply of oxygen to roots can occur in a well-aerated soil (*i.e.*, high air-filled porosity and O_2 concentration) if the roots are surrounded by thick water films. Thus, under such conditions, measurement of soil oxygen movement to plant roots would provide a more complete assessment of soil aeration. Lemon and Erickson introduced the method commonly called ODR (oxygen diffusion rate) to measure oxygen diffusion to a root-like electrode inserted into the soil and enveloped in water films of the soil. The ODR technique has been used by a number of researchers to demonstrate the reduction in oxygen supply in compacted soil.

Limiting soil aeration conditions are often transient and strongly affected by wetting and drying cycles in compacted soil. Therefore, the timing of gas composition and ODR measurements is critical for accurate detection of anaerobiosis. Generally, 200 ng cm^{-2} min^{-1} has been considered the minimum ODR value, below which plant roots will not grow. Some researchers have reported values ranging from 50 to 200 ng cm^{-2} min^{-1} as limiting ODR values for turfgrass root growth. The interpretation of these results has been questioned because, in addition to the diffusion of oxygen, other soil factors and components could have affected the measured ODR.

Work by Blackwell has helped to reduce errors associated with reactions other than the reduction of oxygen in the ODR technique. Moreover, early studies of limiting ODR were performed in growth chambers at about 20æ% C, whereas soil temperatures in the field can vary considerably during the growing season. Blackwell and Wells concluded that the effect of oxygen flux on root growth is related to temperature differences and differences in root respiration. Thus, it is plausible that the oxygen flux limiting root growth under high soil temperatures may be greater than values that have been observed at 20æ% C.

When oxygen becomes limiting, some species of bacteria shift metabolic pathways so as to utilize other compounds as terminal electron acceptors. Thus, plants may have not only to survive periods without oxygen but also to withstand toxic substances such as hydrogen sulfide, which are produced during anaerobic microbial respiration. Development of aerenchyma tissue (intercellular spaces) that allows diffusion of oxygen from aerial plant tissue down to roots may be an adaptive response that helps certain species of plants to survive periods of soil anaerobiosis. Agnew and Carrow reported increased root porosity in Kentucky bluegrass (Poa pratensis L.) grown under compacted soil conditions. They observed the greatest root porosity under conditions of both compaction and water stress.

Soil Strength

Soil strength can be described as the capacity of soil to resist a force without rupture, fragmentation, or flow. Various methods exist to quantitate soil strength in the laboratory, but the method most used in the field is the penetrometer. It measures the effort necessary to push a thin, cone-tipped probe into the soil. Taylor and colleagues conducted pioneering research demonstrating that the strength of a soil, as measured with a static penetrometer, will increase as soil bulk density increases and water content decreases; consequently, mechanical impedance to root growth is greater. The limiting strength of soil to cotton (Gossypium hirsutum L.) root growth is in the range 2.5 to 3 MPa. For many crops, a 2-MPa penetrometer resistance is commonly encountered under field conditions and can reduce root length and elongation by at least 50%.

Root Morphology

In experiments designed to study the constricted growth of roots, the ability of ryegrass (Lolium perenne L.) roots to enter rigid pores (glass capillaries and sheets of steel) depended on the size of the root cap and the thickness of the stele, which needed to be about one-third of the nominal root thickness. Roots could elongate down long capillaries while constricted, albeit at a reduced rate. Roots can enter pores of diameter smaller than the root tip itself only if the rigidity of the pore structure is weak enough to allow soil displacement. Thus, major root axes typically thicken when soil pores are too small for root penetration. Ethylene production has been linked to this thickening response of roots subjected to mechanical impedance. Abscisic acid, which increases temporarily under conditions of mechanical impedance, has also been shown to induce root thickening as well as increased root-hair number and curling of roots. The thickening of the root compresses soil laterally and minimizes friction with soil as the root extends axially. Additionally, root-hair development aids in anchoring the root as it penetrates compacted soil.

Thus these responses can be viewed as adaptive, as they mitigate the effects of axial resistance on growth. Additional contributing factors to thicker roots' relative endurance of compact soil might be greater resistance of thicker roots to bending or higher axial pressures exerted by thick roots. Variation in plant genetics exists for tolerance to high mechanical impedance. Interspecific comparisons have shown differences in root dimensions and the ability to penetrate soils. Greater root thickness and the tendency for roots to expand radially in response to mechanical impedance are correlated with the capacity to elongate in hard soil. Intraspecific studies have also demonstrated genetic variation in characteristics associated with tolerance to mechanical impedance. Two tall fescue (Festuca arundinacea Schreb.) lines with large-diameter roots were able to penetrate a hardpan at the 0.4 to 0.6 m soil depth, extracted more soil water from the 0.6 to 1.2 m depth, and yielded 40% more in dry matter compared to fescues with small-diameter roots.

Development of efficient techniques to screen plant germplasm for greater tolerance to high mechanical impedance could prove to be highly useful. Iijima *et al.* described the morphological development of rice (Oryza sativa L.) and maize (Zea mays L.) root systems as affected by soil compaction. Elongation of the main root axes of rice and maize was restricted in compacted soil. Seminal roots of maize were much less restricted than those of rice. The root system of rice was characterized by a larger fraction of long, thick laterals with potential to produce higher-order laterals on their axes. Growth of higher (second- and third-) order lateral roots in compacted soil compensated for the restricted growth of the main root axes in both species. The work of Carrow and colleagues indicates that the effects of compacted soil on root growth varies with plant species and management. Total root mass of Kentucky bluegrass decreased in response to compaction treatment, whereas tall fescue and perennial ryegrass root systems were more tolerant to compaction stress.

Subsequent work indicated that the most detrimental effect of compaction on root growth of tall fescue and perennial ryegrass occurred under higher nitrogen fertilization. Although total root mass of a turf may not change under compacted conditions, a redistribution of the root system can occur where root mass increases in the surface 5 cm of the soil at the expense of rooting below the 10-cm depth. Root mass may not provide the most detailed index of a root system's response to compacted soil, since root mass is affected less by bulk density than root length is, primarily because of larger-diameter roots in high-density soil. Thus, the measurement of the number and length of roots growing in soil may be a more sensitive indicator of compaction stress on a root system than root mass. Root length and number measurements, however, are expensive, particularly for experiments evaluating a large number of treatments and for plant systems such as turf, which have extremely high root lengths.

Physical alteration of water and nutrient uptake by a root system can result as the orientation and morphology of roots change. Higher soil bulk density brings more soil, and associated water and nutrients, into contact with the surface area of the root system and thus has the potential to increase availability. This effect is particularly pronounced for nutrients such as phosphorus, which have relatively low mobility in the soil. Greater soil contact with the root system, and thus availability of nutrients and water, could at least partly explain the greater clipping yields of turf growing on heavily compacted plots compared to lightly compacted plots during the spring. The fact that this response was reversed during the summer suggests that the beneficial response of greater soil and root surface contact may be beneficial until environmental conditions become limiting for other essential growth factors under compacted soil conditions. As environmental conditions become more limiting, the reduced total root elongation rate and length in compacted soil ultimately decreases availability of water and nutrients by restricted access.

Shoot Responses

The direct and indirect effects of compaction on plant roots can elicit responses in the plant shoot as well. A number of studies on grasses demonstrate the potential of compacted soil to limit shoot density, verdure, clipping yield, soil cover, and lateral stem growth. Lower dry-matter production and yield under compacted soil conditions is attributed to reduced light interception caused by restricted leaf area development and is not a result of an impaired ability of crops to utilize intercepted radiant energy. Hormonal mechanisms have been proposed through which roots "sense" the mechanical impedance of the soil and elicit anatomical changes in leaf growth. An increase in abscisic acid in the xylem sap appears to be a root-to-shoot signal that lowers stomatal conductance and leaf expansion rates.

4

The Production of Cells

Moving fluids such as waterfalls, fountains, and the wakes of boats can generate specific forms. The study of the motion of fluid particles and the shape changes that the fluids undergo is called kinematics. The ideas and numerical methods used to study these fluid forms are useful for characterizing meristematic growth. In both cases, an unchanging form is produced even though it is composed of moving and changing elements.

An example of an unchanging form composed of changing and displaced elements in plants is the hypocotyl hook of a dicot, such as the common bean. As the bean seedling emerges from the seed coat, the apical end of the hypocotyl bends back on itself to form a hook. The hook is thought to protect the seedling apex from damage during growth through the soil. During seedling growth (in soil or dim light) the hook migrates up the stem, from the hypocotyl into the epicotyl and then to the first and second internodes, but the form of the hook remains constant.

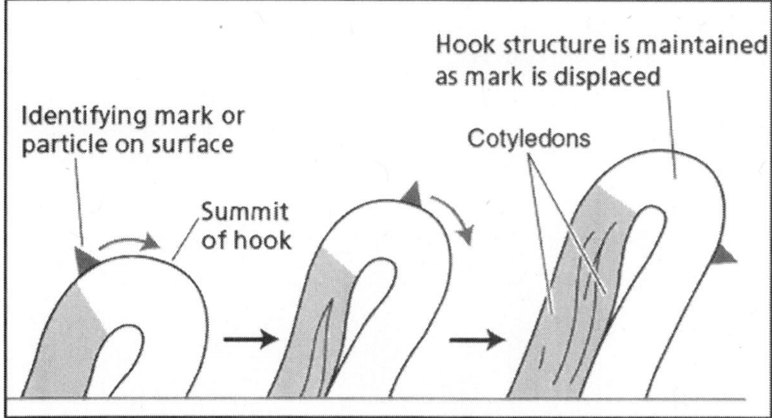

Fig. The dicot Hypocotyl hook is an example of a Constant form Composed of changing Elements.

The hooked form is maintained over time, while different tissues first curve and then straighten as they are displaced from the seedling apex during growth.

If a mark is placed at a fixed point on the surface, it will be displaced (indicated by the arrow), appearing to flow through the hook over time. If we mark a specific epidermal cell on the seedling stem located close to the seedling apex, we can watch it as it flows into the hook summit, then down into the straight region below the hook. The mark is not crawling over the plant surface, of course; plant cells are cemented together and do not experience much relative motion during development. The change in position of the mark relative to the hook implies that the hook is composed of a procession of tissue elements, each of which first curves and then straightens as it is displaced from the plant apex during growth. The steady form is produced by a parade of changing cells.A root tip is another example of a steady form composed of changing tissue elements. Here, too, the form is observed to be steady only when distance is measured from the root tip. A region of cell division occupies perhaps 2 mm of the root tip.

The elongation zone extends for about 10 mm behind the root tip. Phloem differentiation is first observed beginning at 3 mm from the tip, and functional xylem elements may be seen at about 12 mm from the tip. A marked cell near the tip will seem to flow first through the region of cell division, then through the elongation zone and into the region of xylem differentiation, and so on. This shifting implies that developing tissue elements first divide and elongate, and then differentiate.In an analogous fashion, the shoot bears a succession of leaves of different developmental stages. During a period of 24 hours, a leaf may grow to the same size, shape, and biochemical composition that its neighbour had a day earlier. Thus, shoot form is also produced by a parade of changing elements that can be analyzed with kinematics. Such an analysis is not merely descriptive; it permits calculations of the growth and biosynthetic rates of individual tissue elements (cells) within a dynamic structure.

TISSUE ELEMENTS ARE DISPLACED DURING EXPANSION

As we have seen, growth in shoots and roots is localized in regions at the tips of these organs. Regions with expanding tissue are called growth zones. With time, meristems move away from the plant base by the growth of the cells in the growth zone.

If successive marks are placed on the stem or root, the distance between the marks will change, depending on where they are within the growth zone. In addition, all of these marks will move away from the tip of the root or shoot, but their rate of movement will differ depending on their distance from the tip. From another perspective, if you were to stand at the tip of a root that had marks placed at intervals along the axis, you would see that all marks would move farther away from you with time. The reason is that discrete regions on the plant axis experience displacement as well as expansion during growth and development.As a given region of the plant axis moves away from the apex, its growth velocity increases (the rate of elongation accelerates) until a constant limiting velocity is reached equal to the overall organ extension rate. The reason for this increase in growth velocity is that with time, progressively more tissue is located between the moving particle and the apex, and progressively more cells are expanding, so the particle is displaced more and more rapidly. In a rapidly growing maize root, a tissue element takes about 8 hours to move from 2 mm (the end of the meristematic zone) to 12 mm (the end of the elongation zone).

Beyond the growth zone, elements do not separate; neighbouring elements have the same velocity (expressed as the change in distance from the tip per unit of time), and the rate at which particles are displaced from the tip is the same as the rate at which the tip moves through the soil. The root tip of maize is pushed through the soil at 3 mm h^{-1}. This is also the rate at which the Non-growing region recedes from the apex, and it is equal to the final slope of the growth trajectory.

THE GROWTH OF VELOCITY

The velocities of different tissue elements are plotted against their distance from the apex to give the spatial pattern of growth velocity, or growth velocity profile. Growth velocity increases with position in the growth zone. A constant value is obtained at the base of the growth zone.

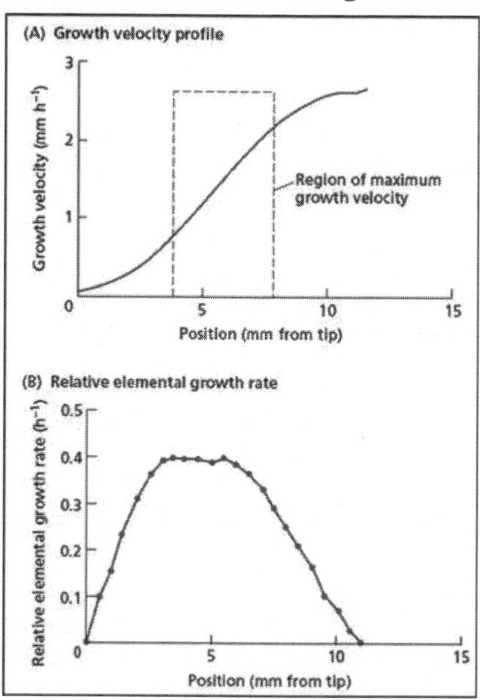

Fig. The Growth of the Primary root of Zea mays (maize) can be represented Kinematically by two related growth curves.

(A) The growth velocity profile plots the velocity of movement away from the tip of points at different distances from the tip. This tells us that growth velocity increases with distance from the tip until it reaches a uniform velocity equal to the rate of elongation of the root. (B) The relative elemental growth rate tells us the rate of expansion of any particular point on the root. It is the most useful measure for the physiologist because it tells us where the most rapidly expanding regions are located.

The final growth velocity is the final, constant slope of the growth trajectory equal to the elongation rate of the organ. In the rapidly growing maize root, the growth velocity is 1 mm h^{-1} at 4 mm, and it reaches its final value of nearly 3 mm h^{-1} at 12 mm.If the growth velocity is known, the relative elemental growth rate, which represents the fractional change in length per unit time, can be calculated. The relative elemental growth rate shows the location and magnitude of the extension rate and can be used to quantify the effects of environmental variation on the growth pattern.

The Relative Elemental Growth Rate

One way to characterize the growth pattern along an axis mathematically is to plot the distance of a tissue element from the apex versus time. The resulting curve is called the growth trajectory. The slope of the curve at any given point is equivalent to the growth velocity at that region of the axis. With the apex as the reference point, the growth trajectory is concave. Marked tissue elements at different locations may be followed simultaneously to generate a family of growth trajectories.

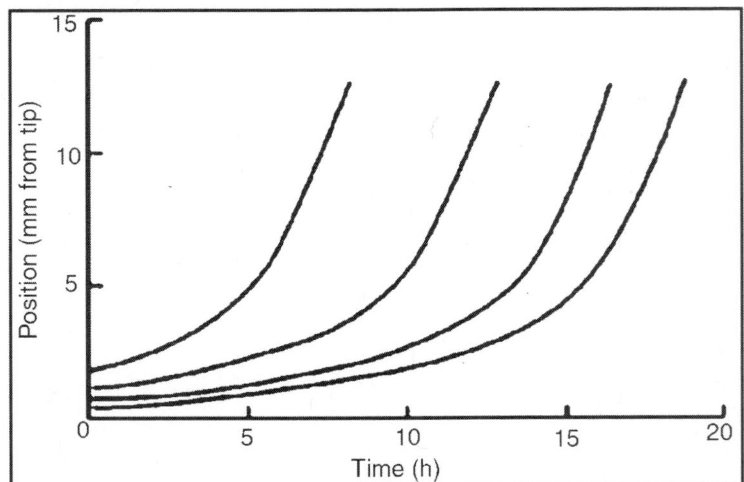

Fig. Growth Trajectories of *Zea mays* (maize): Growth Trajectories plot the positions of points spaced along the root axis as a Function of time.

Note that distance is measured from the root tip. This is termed a material specification of growth because real or material particles are followed. As the regions of the plant axis move away from the apex, their growth velocity increases (*i.e.*, the cells accelerate) until a constant limiting velocity is reached equal to the overall organ extension rate. The reason for this increase in growth velocity is that with time, progressively more tissue is located between the moving particle and the apex, and progressively more cells are expanding, so the particle is displaced more and more rapidly. In a rapidly growing maize root, a tissue element takes about 8 hours to move from 2 mm (the end of the meristem) to 12 mm (the end of the elongation zone). The root tissue element accelerates through this region. Beyond the growth zone, elements do not separate; neighbouring elements have the same velocity, and the rate at which particles are displaced from the tip is the same as the rate at which the tip is pushed away from the soil surface.

The root tip of maize is pushed through the soil at 3 mm h^{-1}. This is also the rate at which the Non-growing region recedes from the apex, and it is equal to the final slope of the growth trajectory. The growth trajectory is a material description of growth because an individual element is followed through time. The growth trajectory provides a way to convert spatial information into a developmental time course.

For example, if the amount of suberin in the root increases with position, the growth trajectory will give information on the time required for the observed accumulation of suberin in cells moving between any two spatial points. To characterize the local expansion rate, we must consider the growth velocities at the apical and basal ends of a small segment of tissue. If the apical end of the segment is moving faster than the basal end, the segment is elongating.

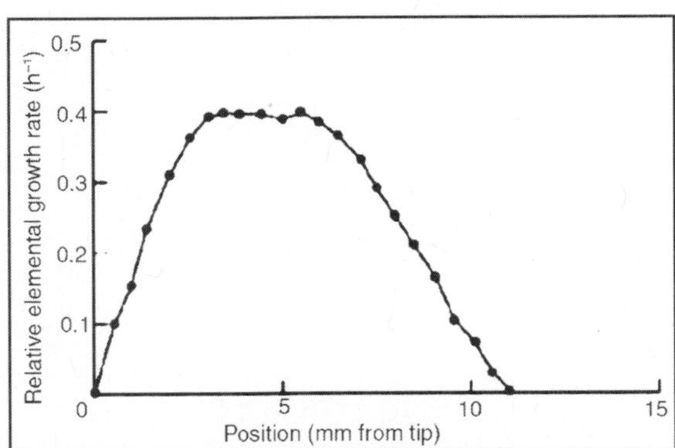

Fig. Relative Elemental Growth rate of *Zea mays* (maize).

The velocity difference, divided by the segment length, gives the relative growth rate of the segment. Now, imagine that the segment shrinks to a point x. To find the relative growth rate, r, at point x, we can use calculus: The value of r is given by the velocity gradient, dv/dx, where v is the growth velocity. This measure of the local growth rate is called the relative elemental growth rate. It represents the fractional change in length per unit time and has units of h^{-1}. If the growth velocity is known, the relative elemental growth rate can be evaluated by differentiation of the growth velocity with respect to position. Or we can approximate r by measuring the relative growth rates of closely spaced segments of initial length L. For each segment, $r = (1/L)(dL/dt)$.

The relative elemental growth rate tells us the rate of expansion of any particular point on the root. It is the most useful measure of root growth rate for the physiologist because it tells where the most rapidly expanding regions are located. The relative elemental growth rate profile, $r(x)$, is a spatial description of the expansion pattern in the growth zone. In the maize root, the relative elemental growth rate has a maximum of 0.40 h^{-1} at 4 mm and drops to zero by 12 mm. That is, the tissue element at 4 mm is expanding 40% h^{-1}, while tissue beyond 12 mm has ceased expansion. Again, to understand the profile in physiological terms, we must consider that individual cells expand more and more rapidly as they are displaced to 4 mm and then less and less rapidly as they are displaced through the basal part of the growth zone.

The relative elemental growth rate profile shows the location and magnitude of the extension rate and can be used to quantify the effects of environmental variation on the growth pattern. It is also the basis for studying the physiology of growth. The graph of $r(x)$ can be compared to the distribution pattern of any substance thought to be regulating the growth rate. A close correlation between the two curves would provide circumstantial evidence for the role of the substance as a limiting factor for growth.

Embryo Maturation Requires Specific Gene Expression

The Arabidopsis embryo enters dormancy after it has generated about 20,000 cells. Dormancy is brought about by the loss of water and a general shutting down of gene transcription and protein synthesis—not only in the embryo, but also throughout the seed. To adapt the cell to the special conditions of dormancy, specific gene expression is required.

For example, the *A*bscisic *A*cid *I*nsensitive*3* (*ABI3*) and *FUSCA3* genes are necessary for the initiation of dormancy and are sensitive to the hormone abscisic acid, which is the signalling molecule that initiates seed and embryo dormancy.

ABI3 also controls the expression of genes encoding the storage proteins that are deposited in the cotyledons during the maturation phase of embryogenesis.The *LE*AFY *C*OTYLEDON*1* (*LEC1*) gene also is active in late embryogenesis. Because *lec1* mutants cannot survive desiccation and do not enter dormancy, the embryos die unless they are rescued through isolation before desiccation occurs. The rescued embryos will germinate in culture and produce fertile plants, which are like wild-type plants except that they lack the 7S storage protein and they have leaflike cotyledons with trichomes on their upper surface. The normal appearance and development of the mature *lec1* mutants indicates that the *LEC1* gene is required only during embryogenesis. Although the most obvious defects of the *lec1* mutants are seen only in the maturation phase embryo, mRNA from *LEC1* gene expression can be detected throughout embryogenesis. It has been proposed that *LEC1* is a general repressor of vegetative development and its expression is necessary throughout embryogenesis.

CELL EXPANSION OF PLANTS

Within 5-10 minutes of application of auxins to isolated epicotyls, or intact tip of the plant or excised segments, growth in their length starts. This early effect is rapid and spontaneous. This rapid growth continues up to 30-45 minutes and then slows down but remains steady. This is also accompanied by an increase in cellular respiration, protein synthesis and the rapid extrusion of protons or in particular Hydrogen ions through the cell walls.

Another significant observation of this early rapid growth is insensitive to the inhibition of transcription (RNA synthesis) and translation (protein synthesis) - ex. Actinomycin-D at high concentration inhibits RNA synthesis completely and cycloheximide (CHI) inhibits cytosol protein synthesis. This clearly suggests that the early rapid growth requires neither the synthesis of RNA nor the synthesis of proteins. But this early effect of auxin is totally inhibited by colchicines at 0.5 to 1% concentration. Colchicine, an alkaloid from the tubers of Colchicum autumnale brings about the inhibition of tubulin (TB) polymerization into long microtubules, which act as the cytoskeleton for mechanical support in the cytoplasm. The colchicines effect on early growth further indicates the important role played by the assembly of microtubules in providing a mechanical structure and the force for all expansion.

Increased metabolic activity, particularly the degradation of osmotically inactive starch etc. leads to the increase of osmotic pressure (they are related to each other) and a steep DPD gradient is created.

This facilitates the inward diffusion of water; consequently turgour pressure builds up within the cells. This force acts on all sides of the cell wall and forces the cell to expand in length and breadth. However recent investigations have shown that the cell elongation takes place under negative pressure.

The above features suggest that the process or the mechanism of cell elongation is complex and multiphased, but highly regulated. Various theories have been proposed from time to time and none of them could prove their claim because of one or the other drawbacks. Here we will restrict to a comprehensive account of cell elongation.

Early phase: The auxin's effect in the early phase on cell growth is rapid. This is insensitive to the inhibitors of transcription and translation but insensitive to colchicines and respiratory inhibitors like cyanide, DNP etc. From this it can be deduced that cell elongation does not require new RNA and protein synthesis, but it requires ATP and ATP dependent tubulin polymerization into microtubular cytoskeleton. Auxin at the early stages activates and increases the rate of respiratory metabolism.With the result ATP pool builds up rapidly. Auxin also activities membrane bound nucleating centre of tubulin polymerization.These nucleating centres, utilizing ATP and tubulin monomers found in the cytoplasm, start polymerizing and soon build up long microtubules. These, in turn, start orienting parallel to the long axis of the cell, which probably requires cytochalasin-B sensitive microfilaments.With the rapid increase in microtubules, oriented in longitudinal plane, considerable force is exerted on the cell boundaries.

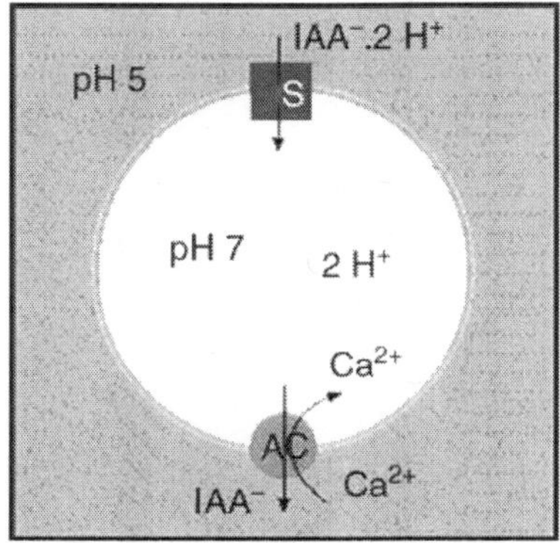

As the microtubular polymerization is building up, ATP ase/H* gets activated, and H* ions are pumped out into the interspaces of cellulose fibrils acidic pH in turn activates cellulases enzymes, which starts loosening up the cell-wall cellulose fibres. This loosening up of cell wall concomitant with acting as a mechanical force-the cell expands in length. This process starts 4-5 minutes after application of auxins and the elongation continues rapidly for about 20-30 minutes.

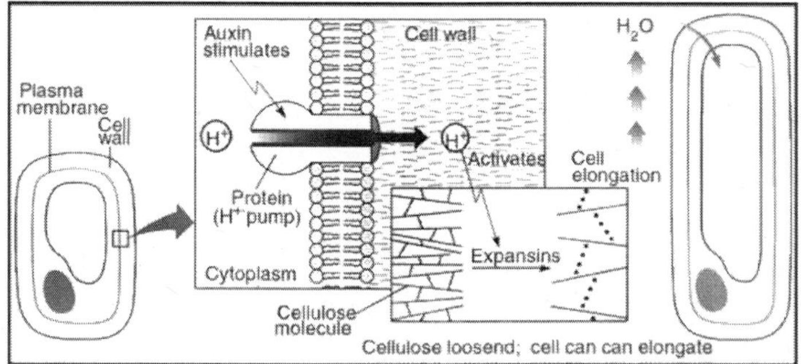

The late phase starts around 30-40 minutes after auxin application. This is slower than the first phase, but growth is steady. This phase is sensitive to Actinomycin-D, CHI, Colchicine and respiratory inhibitors, indicting that this process requires metabolic energy, tubulin-polymerization, RNA synthesis as well as protein synthesis. As tubulin pool gets exhausted in the first phase, it has to be replaced. That means it requires the synthesis of RNA and proteins.

Apart from these proteins, the cytoplasm requires the synthesis of other proteins, for the increase in the mass of protoplasm. It is quite possible that the early activation of early translation may have significant effect on gene activation. Thus it takes some time for the second phase of growth to set in. The second phase provides all materials for sustained growth.

Growth Movements: The phenomenon of bending movement of plants in response to light stimulus is called Phototropism, the growth of roots towards soil is called Geotropism.

The curvature growth due to the touch is named as Thigmotropism etc. Such differential growth movements are due to differential concentration of auxin. The stem tips and root tips have different optima of their auxin requirement to show maximum growth.

The mechanism of these movements will be explained in the chapter on growth movements. Auxin brings about phototropic movement because differential accumulation of auxin at exposed and darker side.

Root initiation: Vegetative propagation of stem cuttings is a normal practice in horticulture, floriculture and agriculture. Out of some stem cuttings which are put into soil, a few produce roots easily but others may or not produce roots. The roots that develop from any part of the plant body other than the radicle are called adventitious roots.

The development of such roots from the stem or any other part is a phenomenon of dedifferentiation. Stem is quite different fro the root with respect to its structure and function.

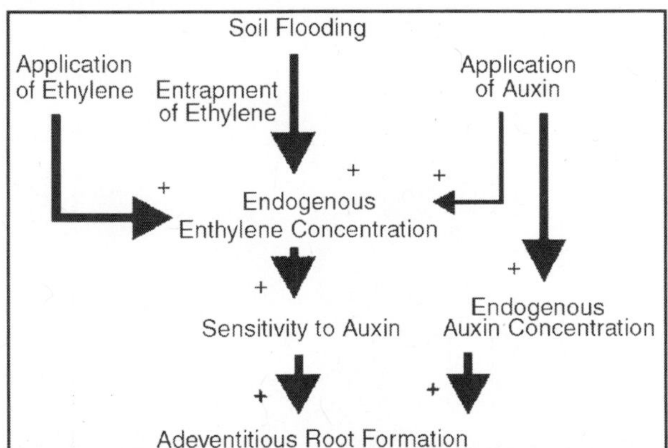

Still some of the tissues found in stem undergo dedifferentiation and develop into roots. This dedifferentiation and growth of roots from the stem is initiated by auxins Indole acetic acid, IBA, NAA, and other hormones. When applied to stem cuttings they induce the formation of roots. IBA has been found to be very good in this regard.This indicates that the pre-existing m RNA protein complex is drawn towards protein synthesis: it shows increased rate of protein synthesis. Some of the proteins thus synthesized in turn activate differential nuclear gene activation. As a result of it, a cascade of events occurs. Generally there is an increase in the synthesis of all cellular proteins. At the same time there is an enhanced rate of synthesis of some specific proteins. Also some new proteins are synthesized. Among these, tubulin, a membrane bound fraction, is synthesized at a high rate. It is believed that this protein along with microfilaments gets assembled into microtubules which orient themselves in a specific way so as to facilitate the polarity fixation for cell division.

In response to auxins, some of the cells in the pericycle region get stimulated and the required macromolecules are synthesized by differential gene activation. Thus these cells endowed with these new potentialities divide and redivide, and soon they organize into root primordial. They in turn develop into new roots which penetrate through cortical cells. The time taken for this new root formation, in response to auxin supply is just 24-36 hours. This effect of auxin induced new root formation has been well exploited by horticulturists and others. The following diagram shows how auxins interact with cytokinins in eliciting growth responses.

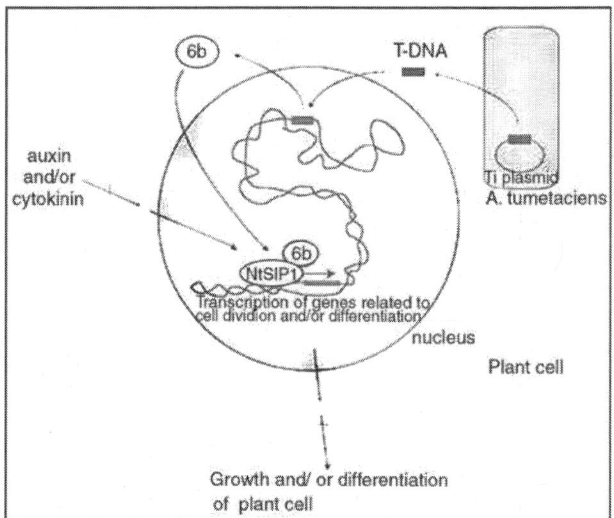

Apical dominance: The active growth of terminal bud inhibits the development of lower axillary bunds. This effect is called apical dominance. Many plants like conifers and other exhibit the apical dominance so predominantly that the plant grows tall with few branches and appear conical in shape. But plants like banyan, mango, tamarind do not show that much of apical dominance, hence these plants spread outwardly. This phenomenon of apical dominance has been attributed to the differential concentration of auxin and nutrient supply to the axillary buds.

Abscission: With the onset of winter season the leaves of deciduous plants turn yellow and fall off. The falling of the leaves is due to the formation of abscission layers at the base of the petioles. This layer consists of thin walled but suberized cells. The middle wall gets disintegrated and the leaves fall down. The development of this layer is stimulated by a plant hormone called ABA. But the application of auxin to such leaves prevent the formation of abscission layer and thus prevent the falling off the leaves. This has a greater application particularly in preventing the premature falling of fruits like apple, orange etc.

Parthenocarpy: Development of fruits without fertilization is referred to as parthenocarpy and such fruits are seedless. Normally plants do not produce parthenocarpic fruits. Normally plants do not produce parthenocarpic fruits.

But the application of auxin to flowers induces parthenocarpic fruit setting and prevents premature falling of fruits as well. Plants with triploid and many polyploid genomes do not set fruits. If such plants are sprayed with auxins at the time of flowering, fruit setting is enhanced. The fruits thus developed are large, seedless, and sweeter. Seedless oranges, lime, watermelon, sapota, ramphal and seethaphal, if they are produced this way, they have good market at the commercial level.

Gibberllic Acid

Gibberllic Acid is another natural hormone found distributed in higher plants. This was first discovered in Japan, later its structural and functional details have been elucidated by the people in the west. So far 52 or more different kinds have been identified. Among them GA2 2GA1 are the most common ones found in all plants.

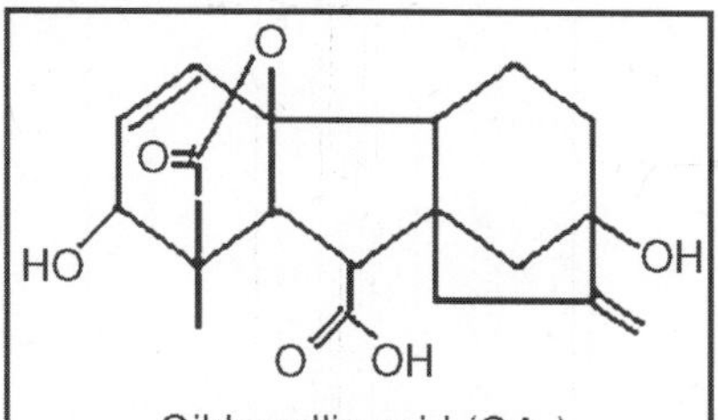

Gibberellic acid (GA_3)

GA is a steroid like molecule in its structure. Young leaves are the sites in which this hormone is mediated by phytochromes and then it is transported to the regions of growth.

The most significant effect of GA on the growth of plants is it effects internodal cell elongation profoundly than any other hormones. Even genetic dwarfs (eg. Dwarf Pisum sativum) grow as tall as the natural tall plants.

It also brings about a dramatic effect called Bolting and Flowering. Long day plants (which require 12-14 hrs of day light treatment for flowering) under short day conditions produce a cluster of leaves on highly shortened stem and never produce flowers. When such plants are sprayed with GA, the stem starts elongation with enhanced growth of internodes. Simultaneously the terminal buds starts producing floral branches. This phenomenon is called bolting and flowering. However both bolting and flowering are separated by time and space.Besides these above said effects, GA is essential for the mobilization of food reserves for the germinating cereal grains like wheat and maize. At the time of germination, the young embryonic axis synthesizes this hormone, which is soon released into endosperm and reaches the aleurone layer. It is this layer, GA activates differential gene expression. As a result, specific m.RNAs is transcribed and later specific proteins like amylase are synthesized on translation. The amylase thus synthesized, acts upon the starch found in the endosperm, and the glucose thus released is drawn into the embryonic tissue which is in dire need of energy source. Notwithstanding the above effects, GA also effects parthenocarpy, rooting, germination etc. This hormone also interacts with other natural hormones like auxin and cytokinin and brings about differential growth.

CYTOKININS

Another class of growth regulating hormones has been discovered by Miller et al and they are called Cytokinins. They are the derivatives or nucleotides. 6-Furfuryl amino purines, Benzyl amino purines and such compounds act as cytokinins.

Cytokinins are generally synthesized in the root tips, and then they are translocated to other regions. The endosperm of coconut milk, Borassus and of other plants is a rich source of cytokinins. Cytokinin level in plants is further regulated by other hormones like Auxins and GAs. On the contrary cytokinins control the synthesis of other hormones like Abscisic acid, a growth inhibitor.

Cytokinin per se has an important role in the cytokinesis. But along with auxins, it effects differentiation of plant organs from the callus. Cytokinin by itself inhibits root formation, but in the presence of auxins, it induces organ formation like roots and shoots. If the ratio between auxin and cytokinin is high, roots are differentiated, but the lower ratio induces the formation of shoots.

The intermediate ratio stimulates the development of these hormones in differentiation at the molecular level is very interesting by unfortunately it is least understood. Cytokinins are also found to break dormancy of winter buds as well seeds. The exact mechanism by means of which they over come the dormancy is not known, still it has been thought, that they effect through the suppression of IBA which is responsible for dormancy. Another interesting effect that the cytokinins produce is the prolongation of senescence (Ageing). This effect is often referred to as Richmond and Lang's effect.

5

Plant Breeding: Concept and Aim

METHODS IN PLANT BREEDING

CONVENTIONAL BREEDING

Conventional plant breeding involves crossing carefully chosen parent plants, then selecting the best plants from the resulting offspring to be grown on for further selection. For cereals, hundreds of individual crosses are carried out by hand to create seed for the first filial (or F1) generation. The resulting F1 plants are uniform, but in the following generation several hundred thousand different plants are produced. Because of the way genes work, the new combinations produced from each cross are not revealed until the second (F2) generation.

It is this enormous diversity of new gene combinations that may hold the key to a successful new variety. The plant breeder's task is to select the plants most likely to meet his breeding objectives. Seed from the best of these F2 plants is grown on in small rows or plots and the best plants again selected– this process is repeated year after year until only the very best plants remain. As promising new lines emerge, tests are conducted on each plot to assess factors such as yield, disease resistance and endues quality.

Once the best lines are purified to ensure that every plant has the same characteristics, the process of multiplying seed begins. These 'inbred' lines are then ready for entry into official trials, some six to ten years after the initial cross.

Hybrid Breeding

For some crop species, the seed supplied to growers is that produced from the first cross between selected parents. The resulting varieties, known as F1 hybrids, offer potential advantages in crop performance. Hybrid breeding is widely used to produce varieties of field vegetables, maize and oilseed rape. F1 hybrids are unique in expressing 'hybrid vigour' in the growing crops for a single year. This may result in higher yields, greater uniformity, or improvements in quality. Unlike inbred lines, F1 hybrids do not breed true year after year and their performance gains are not maintained in subsequent generations.

Improved Breeding

With increasing knowledge and improved technology, breeders have developed ways to enhance the speed, accuracy and scope of the breeding process.There are several ways to reduce the lengthy interval between the first cross of selected parents and establishing true breeding lines of promising new varieties. For example, maintaining parallel selection programmes in northern and southern hemispheres allows two generations to be produced each year. Single seed descent produces very small plants under restricted growth conditions. Large numbers of these plants are cultivated in artificial growth rooms, with two or more generations produced in a year. In potatoes, mini-tuber breeding speeds up the slow multiplication process by producing miniature plants under greenhouse conditions.

More recent laboratory techniques enable breeders to operate at the level of individual cells and their chromosomes. In certain crop species, such as potatoes and oilseed rape, it is possible to produce new varieties in the laboratory through

Protoplast Fusion

Individual plant cells with their outer walls removed are fused, and the fused cells then induced to divide and grow in a culture medium. Whole plants are eventually regenerated containing new combinations of genes from the two parents.

Embryo Rescue and Assisted Pollination

Allow breeders to expand the range of available characters by making crosses between species, which would not produce viable offspring outside the laboratory.

Double Haploid Breeding

Enables breeders to produce genetically uniform lines within one generation. This effectively by-passes the lengthy process of self pollination and selection normally required to produce true breeding plants. Latest developments in genetic science have greatly improved our understanding of how plants behave, offering additional ways to enhance the breeding process.

Genomics

By mapping the genetic makeup, or genome, of crop species, scientists can identify the exact position and function of individual genes. Genome mapping has revealed striking similarities in the genomes of different crop species, such as rice, wheat, barley and rye. This information is already helping to broaden the scope and precision of current breeding programmes.

Marker Assisted Breeding

Allows breeders to determine whether desired traits are present in a new variety at an early stage in the breeding programme.

Genetic Modification

Genetic modification of plants is achieved by adding a specific gene or genes to a plant, or by knocking out a gene with RNAi, to produce a desirable phenotype. The resulting plants are often referred to as transgenic plants. Genetic modification can produce a plant with the desired trait or traits faster than classical breeding because the majority of the plant's genome is not altered.

To genetically modify a plant, a genetic construct must be designed so that the gene to be added or knocked-out will be expressed by the plant. To do this, a promoter to drive transcription and a termination sequence to stop transcription of the new gene, and the gene of genes of interest must be introduced to the plant. A marker for the selection of transformed plants is also included. In the laboratory, antibiotic resistance is a commonly used marker: plants that have been successfully transformed will grow on media containing antibiotics; plants that have not been transformed will die. In some instances markers for selection are removed by backcrossing with the parent plant prior to commercial release.

The construct can be inserted in the plant genome by genetic recombination using the bacteria Agrobacterium tumefaciens or *A. rhizogenes*, or by direct methods like the gene gun or microinjection. Using plant viruses to insert genetic constructs into plants is also a possibility, but the technique is limited by the host range of the virus. For example, Cauliflower mosaic virus (CaMV) only infects cauliflower and related species. Another limitation of viral vectors is that the virus is not usually passed on the progeny, so every plant has to be inoculated.

The majority of commercially released transgenic plants, are currently limited to plants that have introduced resistance to insect pests and herbicides. Insect resistance is achieved through incorporation of a gene from Bacillus thuringiensis (Bt) that encodes a protein that is toxic to some insects. For example, if cotton pest the cotton bollworm feeds on Bt cotton it will ingest the toxin and die. Herbicides usually work by binding to certain plant enzymes and inhibiting their action. The enzymes that the herbicide inhibits are known as the herbicides *target site*. Herbicide resistance can be engineered into crops by expressing a version of *target site* protein that is not inhibited by the herbicide. This is the method used to produce glyphosate resistant crop plants.Genetic modification of plants that can produce pharmaceuticals (and industrial chemicals), sometimes called *pharmacrops*, is a rather radical new area of plant breeding

Proteomics

Allows breeders to understand how genes behave in different parts of the plant and under different growing conditions.

Maintaining Genetic Diversity

Maintaining biodiversity is central to the process of crop improvement. It is in every breeder's interest to ensure that the gene pool from which new traits are selected remains as extensive as possible. Plant breeders created the first gene banks in the 1930s to conserve the valuable genetic diversity within past and present varieties, as well as landraces and wild relatives of cultivated crop species. Plant breeding is integral to ongoing initiatives to identify, classify and conserve existing biodiversity.

Testing Plant Varieties

Before any new crop variety can be placed on the market, it must undergo statutory testing under a process known as National Listing. Successful varieties are placed on a National List or register of varieties approved for marketing.National Listing rules are determined on a European basis, and apply to all the major agricultural and vegetable crop species. Official trials are conducted, in most cases for a minimum of two years, to test each new variety for a range of characteristics which together determine its uniqueness, its genetic uniformity, and its value to growers and the rest of the food chain. National Listing is extremely rigorous–the majority of varieties entered do not complete the process. In winter wheat, for example, only a quarter of varieties entered for National Listing during the 1990s were finally approved.

National Listing–DUS

All varieties entered for National Listing are assessed for Distinctness, Uniformity and Stability (DUS). In the case of cereals, some 30 individual characteristics of the plant are inspected to verify that it is distinct, ie clearly distinguishable from other varieties, that its characteristics are uniform from one plant to another, and that the variety is stable in that it breeds true to type from one generation to the next.

National Listing–VCU

For agricultural crops, National Listing also involves trials to establish a variety's Value for Cultivation and Use (VCU). This provides an assurance to growers that only varieties with improved performance or end-use quality can be officially approved for sale.

Recommended and Descriptive Lists

Once a new variety has been added to the National List, it is cleared for marketing. However, its success in the market place is by no means guaranteed. Further independent trials are conducted each year to compare the performance and quality of the best varieties. These trials provide the basis for detailed information and advice to growers and their customers.

As shown in the following illustration for winter wheat, the process of statutory and commercial evaluation can take up to five years. Only a handful of varieties clear all these hurdles, which are in addition to the many years spent testing and selecting in the breeder's own trials programme.

Genetically Modified Varieties

No GM crops can be marketed until they have been assessed and approved in terms of human health, food safety and the environment. The cross-border movement of genetically modified organisms (GMOs) is regulated at a global level under the internationally agreed Biosafety Protocol.

Variety Maintenance and Identity Preservation

As well as developing new varieties, plant breeders maintain the genetic purity of existing lines and pre-commercial seed supplies year by year. This process is costly and time-consuming, but essential to maintain the quality and performance of each variety.For cereals, variety maintenance begins after a few years of selection trials, when the promise of a variety is just emerging. At that stage, all that exists of what may become a widely grown variety is a single row containing around 100 plants. The breeder then bulks up supplies of the purified lines of breeder's seed into prebasic and then basic seed. Each year specialist seed growers are used to grow basic seed for the first generation of certified seed–or C1 seed. After one more year this becomes C2 seed, the main source of certified seed used by farmers.

The plant breeder continuously maintains breeder's seed for the process of multiplication through pre-basic, basic, C1 and C2 seed to ensure the variety's performance and quality year after year. Greater emphasis is now being placed on preserving the identity of individual varieties after harvest, both to conserve quality characteristics and to meet consumer demands for assurances about the integrity and traceability of their food.

Seed Certification

Seed of an approved variety can only be marketed if it meets strict quality criteria Seed quality standards are laid down in UK and EU law, and policed by agencies appointed by Government. The UK's official seed certification system offers an independent assurance of quality to growers. Minimum standards apply for varietal identity, purity and germination capacity. In addition, strict limits apply to seed-borne diseases, and the presence of physical impurities such as weed seeds.Around 9% of the UK arable area is used to multiply the pure lines of seed from the plant breeder into certified seed. Several thousand individual crops are involved; each grown under specific management regimes to ensure the purity and integrity of the resulting seed is maintained. To gain certification, every seed crop must undergo crop inspection and seed testing. Seed certification underpins the health and purity status of the major arable crops in Britain. It offers an independent benchmark of quality on which buyers of seed and their customers depend.

Farm-saved Seed

For certain crop species–particularly small-grain cereals–growers can opt to save their own seed for sowing the following year provided care is taken to ensure that the crop remains healthy and free from impurities, and that the resulting seed is carefully conditioned and cleaned. Without independent testing for germination and freedom from seed borne diseases, however, farm-saved seed can harbors risks to growers that may not become apparent until well into the growing season. Many growers choose certified seed for the peace of mind that quality and performance is independently assured.

Plant Breeding in the Future

Success in any line of business depends on being able to anticipate and respond to changing market requirements. Such is the long-term nature of developing a new crop variety that plant breeders are in the unenviable position of forecasting their customers' demands at least five, and probably nearer ten years, in advance. Plant breeding objectives will continue to be shaped by changes in agricultural and environmental policy. There is no doubt that while improvements in crop yield and endues quality remain of paramount importance, the future for British agriculture lies in crop production systems which deliver both economic and environmental benefits.

Increasingly, plant breeders are targeting varieties suitable for low input and organic regimes. It is already estimated that genetic improvements in disease resistance alone save British farmers more than £100 million a year in reduced agrochemical use and improved yield. A range of other factors also determines a variety's suitability for low input, integrated farming systems. They include the crop growth cycle, straw strength and susceptibility to weed competition, as well as the variety's ability to thrive in a range of soil types, climatic conditions and rotational patterns.

Novel Crops

Breeders continue to adapt novel crops for UK growing conditions. New varieties of sunflowers and soya beans are already creating market opportunities in the food sector. Post-BSE, attention has focused on using more home-grown fodder in animal feed. For plant breeders, this means enhancing energy and protein levels in traditional forage crops, and developing the potential of new crop species, such as lupins.

Non-food Crops

The use of crop plants for industrial, Non-food applications is an area of increasing interest. Developments in plant breeding could underpin a transition from many crude, industrial processes towards renewable, field-based production of specialist chemicals, textiles and biofuels Quick-growing biomass crops such as willow coppice and miscanthus are already fuelling UK power stations, and there is huge potential to develop oilseed, starch and fibre crops for a wide range of product including lubricants, pharmaceuticals, dyes, plastics, flooring, paper and clothing.

New Traits

Pioneering research by British scientists is opening up new possibilities for crop improvement. Examples from current research include the use of genetic modification to overcome grey mould infestation in strawberries–not possible through conventional breeding because no resistance genes exist in the world strawberry germplasm. Similar techniques offer ways to control the ripening process in fruit and vegetables, to extend storage life, and improve flavour and texture for consumers. It will also increase the seasonal availability of homegrown produce, so reducing the distance food is transported. Modern biotechnology can be used to improve human nutrition. Recent research by British scientists includes the development of vegetables containing increased levels of cancer-fighting nutrients, and apples, which really do keep the dentist at, bay by protecting against tooth decay.

Patents

Recent developments in technology have also seen the emergence of patent applications in relation to plant breeding. This is particularly the case where technologies can be applied to several different varieties of the same crop, or across a range of different species. Such developments cannot be protected under the variety-specific system of Plant Breeders' Rights.

The use of patents where living organisms are concerned continues to arouse controversy, but future research will not take place unless some mechanism exists to reward and promote sharing of knowledge.

Issues and concerns

Modern plant breeding, whether classical or through genetic engineering, comes with issues of concern, particularly with regard to food crops. The question of whether breeding can have a negative effect on nutritional value is central in this respect. Although relatively little direct research in this area has been done, there are scientific indications that, by favouring certain aspects of a plant's development, other aspects may be retarded. Reductions in calcium, phosphorus, iron and ascorbic acid were also found. The study, conducted at the Biochemical Institute, University of Texas at Austin, concluded in summary: *"We suggest that any real declines are generally most easily explained by changes in cultivated varieties between 1950 and 1999, in which there may be trade-offs between yield and nutrient content*

The debate surrounding genetic modification of plants is huge, encompassing the ecological impact of genetically modified plants and the safety of genetically modified food.

Plant breeders' rights is also a major and controversial issue. Today, production of new varieties is dominated by commercial plant breeders, who seek to protect their work and collect royalties through national and international agreements based in intellectual property rights. The range of related issues is complex. In the simplest terms, critics of the increasingly restrictive regulations argue that, through a combination of technical and economic pressures, commercial breeders are reducing biodiversity and significantly constraining individuals (such as farmers) from developing and trading seed on a regional level. Efforts to strengthen breeders' rights, for example, by lengthening periods of variety protection, are ongoing

Plant Breeding For A World Community

There is enormous potential for plant breeding to benefit less developed parts of the world, where food security is critical as populations continue to increase.

In particular, advances in genetic technology may offer more versatile solutions than conventional systems of crop improvement. For example, scientists in Britain are pioneering important research to develop salt and drought tolerance in crop plants. This may help to alleviate the devastating effects of crop failure in arid regions such as sub-Saharan Africa. Private and public sector initiatives to share knowledge and expertise are already in place. They include commitments to provide Vitamin A enhanced rice and virus resistance in sweet potatoes free of charge to developing countries.

International arrangements to recognize genetic resources, as the sovereign property of individual nation states will stimulate further programmes of technology transfer and benefit sharing between industrialized and less developed regions of the world.

Effects of Modern Plant Breeding on Environment

Some environmentalists are concerned that genes from genetically modified crops could escape and transfer to other species with unwanted consequences. For example, it is argued that herbicide-resistant crops could cross with weedy relatives to create a new strain of 'super weed'. In practice, since the reproductive systems of genetically modified and conventionally bred crops are identical, the behaviour of domesticated crop plants is unlikely to be affected by single gene changes.

In ten years of worldwide field trials there have been no adverse reports of genetically modified crops spreading in the environment. Furthermore, resistance to weed killers has been available for decades in a number of conventionally bred varieties without causing any such problem. The development conventionally bred crops-nevertheless the technology is subject to much tighter controls. In the long history of plant breeding, the strict regulations applied to the development and use of genetically modified crops is unprecedented. In both the laboratory and the field, each genetically modified crop must go through a rigorous process of monitoring and evaluation before it can reach the customer. Extensive and ongoing evaluation of genetically modified crops has shown that the technology presents no new food safety risks. Indeed, biotechnology will enable breeders to develop food crops with improved nutritional value and better keeping qualities. Enhanced disease and pest resistance will also reduce pesticide residues.

CONCEPT

The constant aim of plant breeding is to improve the quality, diversity and performance of agricultural and horticultural crops. The overriding objective is to develop plants better adapted to human needs. The evolution of plant breeding is a classic example of how improved biological understanding has been adapted to provide more effective methods of meeting the demands of a changing world. Plant breeding has existed in its most primitive form since the first farmers saved the seeds of their best plants from one season to the next more than 10,000 years ago. Over the centuries, this selection process has gradually become more scientific, bringing major improvements in the yield, quality and diversity of crops grown worldwide. In the 19th century Gregor Mendel established the basic principles of plant genetics. Gregor Mendel in the 19th century laid the foundations for traditional plant breeding, in which selected parents are cross-pollinated to combine certain desired characteristics, such as high yield and resistance to pests and disease. He discovered that units of material, which are transferred from one generation to the next, determine inherited traits.

The plant breeder's aim is to reassemble these units of inheritance, known as genes, to produce crops with improved characteristics. In practice, this is a complex and time-consuming process. Each plant contains many thousands of genes, and the plant breeder is seeking to combine a range of desirable traits in one plant to produce a successful variety. Conventional breeding involves crossing selected parent plants, chosen because they have desirable characteristics such as high yield or disease resistance. The breeder's skill lies in selecting the best plants from the many and varied offspring. These are grown on and tested in subsequent years. Typically this involves examining thousands of individual plants for different characteristics ranging from agronomic performance to end-use quality.

Developing a new variety can take up to 15 years for wheat, 18 years for potatoes, even longer for some crops. The scope of conventional plant breeding has increased with improvements in technology. In the laboratory, chemical and mechanical techniques are used to speed up the selection process and remove natural barriers to cross-fertilization, for example between different crop species.

All forms of plant breeding involve the improvement of crops by selection. Since man began farming he has selected seeds from the best plants for the next generation. Over the centuries, this selection process has become more scientific, bringing major improvements in the yield, quality and diversity of agricultural and horticultural crops. Genes-units of hereditary material that are transferred from one generation to the next, determine these characteristics. Since the plant contains many thousands of genes, and the breeder is seeking to combine a range of traits in one plant, developing a successful variety can be an extremely lengthy process-up to 12 years in the case of cereals.

Using biological knowledge derived from cross-pollination, plant breeders have developed physical and chemically assisted ways of enhancing the speed, accuracy and scope of the selection process. Breeding systems involving grafting and hybridization, for example, have extended breeders' capabilities at the whole crop level, while more recent cell-based techniques enable breeders to operate at the level of individual cells and their chromosomesA major objective in modern plant breeding is the making of crop varieties with the highest possible yield potential. Yield potential is defined as "the yield of a crop when growth is not limited by water or nutrients, pests, diseases, or weeds". For farmers, whose crops are indeed limited by such constraints, this yield concept may not be seen as the most relevant objective. Relieving crops of all sorts of environmental stress, leaves light and temperature, together with varietal characteristics, as the only determinants of yield. That achieved, the same high yielding variety can be used all over a climatic zone, and target area for the variety is enormous. If stress can not be eliminated, however, adaptation to a usually site-specific environment becomes necessary. In that case, the target area for each variety will be small.

Do these different perspectives warrant different approaches to plant breeding? Conventional wisdom says no. In experiments where varieties are tested at various levels of inputs, the same high yielding variety usually comes out as top yielder at all levels. Therefore, plant breeders often claim that their varieties not only perform excellently under high-input conditions, but would also be better than traditional varieties in a low-input environment. However, when these trials are taken outside of experimental farms and the varieties are tested under local or farm conditions, researchers discover what is called 'crossover in performance'. At a certain level of stress there is a crossover point beyond which local varieties or landraces perform better than the high yielding varieties Farmers who experience such situations are not a tiny minority.

To some degree, most if not all farmers have to cope with local stress conditions. The stress-free environment is hard to achieve, and even harder to sustain. In many favourable areas, farmers abandon the high-input technology for economic reasons or because of ecological problems, thus increasing the need for locally-adapted germplasm. But modern plant breeding, whether public or private, can not supply adapted germplasm everywhere. Only a system of local seed selection can ensure that. And that means devolution of plant breeding.

Can such decentralized breeding be compatible with development needs in a changing world and meet economic aspirations in a poor society? If the answer to these questions is going to be 'yes', the decentralized breeding must be able to take advantage of the power of science as well as of the capacities of local communities. Three aspects should be considered and made mutually compatible: breeding technology, participatory research methods, and organization at community level.

History

Plant breeding is the purposeful manipulation of plant species in order to create desired genotypes and phenotypes for specific purposes. This manipulation involves either controlled pollination, genetic engineering, or both, followed by artificial selection of progeny. *Plant breeding* often, but not always, leads to plant domestication.Plant breeding has been practiced for thousands of years, since near the beginning of human civilization. It is now practiced worldwide by government institutions and commercial enterprises. International development agencies believe that breeding new crops is important for ensuring food security and developing practices of sustainable agriculture through the development of crops suitable for their environment.

Domestication

Domestication of plants is an artificial selection process conducted by humans to produce plants that have fewer undesireable traits of wild plants, and which renders them dependent on artificial (usually enhanced) environments for their continued existence. The practice is estimated to date back 9,000-11,000 years. Many crops in present day cultivation are the result of domestication in ancient times, about 5,000 years ago in the Old World and 3,000 years ago in the New World. In the Neolithic period, domestication took a minimum of 1,000 years and a maximum of 7,000 years. Today, all of our principal food crops come from domesticated varieties.

A cultivated crop species that has evolved from wild populations due to selective pressures from traditional farmers is called a landrace. Landraces, which can be the result of natural forces or domestication, are plants (or animals) that are ideally suited to a particular region or environment. An example are the landraces of rice, *Oryza sativa* subspecies *indica*, which was developed in South Asia, and *Oryza sativa* subspecies *japonica*, which was developed in China.

Classical plant breeding uses deliberate interbreeding (crossing) of closely or distantly related species to produce new crops with desirable properties. Plants are crossed to introduce traits/genes from one species into a new genetic background. For example, a mildew resistant pea may be crossed with a high-yielding but susceptible pea, the goal of the cross being to introduce mildew resistance without losing the high-yield characteristics. Progeny from the cross would then be crossed with the high-yielding parent to ensure that the progeny were most like the high-yielding parent, (backcrossing), the progeny from that cross would be tested for yield and mildew resistance and high-yielding resistant plants would be further developed. Plants may also be crossed with themselves to produce inbred varieties for breeding.

Classical breeding relies on homologous recombination of two genomes to generate genetic diversity. The Classical plant breeder may also makes use of a number of *in vitro* techniques such as protoplas fusion, embryo rescue or mutagenisis to generate diversity and produce plants that would not exist in nature.

Traits that breeders' have tried to incorporate into crop plants in the last 100 years include:

- Increased quality and yield of the crop
- Increased tolerance of environmental pressures (salinity, extreme temperature, drought)
- Resistance to viruses, fungi and bacteria
- Increased tolerance to insect pests
- Increased tolerance of herbicides

Before World War II

Intraspecific hybridization within a plant species was demonstrated by Charles Darwin and Gregor Mendel, and was further developed by geneticists and plant breeders. In the early 20th century, plant breeders realised that Mendel's findings on the non-random nature of inheritance could be applied to seedling populations produced through deliberate pollinations to predict the frequencies of different types.In 1908, George Harrison Shull described heterosis, also known as hybrid vigour. Heterosis describes the tendency of the progeny of a specific cross to outperform both parents. The detection of the usefulness of heterosis for plant breeding has lead to the development of inbred lines that reveal a heterotic yield advantage when they are crossed. Maize was the first species where heterosis was widely used to produce hybrids.

By the 1920s, statistical methods were developed to analyse gene action and distinguish heritable variation from variation caused by environment. In 1933, another important breeding technique, cytoplasmic male sterility (CMS), developed in maize, was described by Marcus Morton Rhoades. CMS is a maternally inherited trait that makes the plant produce sterile pollen, enabling the production of hybrids and removing the need for detasseling maize plants.

These early breeding techniques resulted in large yield increase in the United States in the early 20th century. Similar yield increases were not produced elsewhere until after World War II, the Green Revolution increased crop production in the developing world in the 1960s.

After World War II

Following World War II a number of techniques were developed that allowed plant breeders to hybridize distantly related species, and artificially induce genetic diversity.

When distantly related species are crossed, plant breeders make use of a number of plant tissue culture techniques to produce progeny from other wise fruitless mating. Interspecific and intergeneric hybrids are produced from a cross of related species or genera that do not normally sexually reproduce with each other. These crosses are referred to as *Wide crosses*. The cereal triticale is a wheat and rye hybrid. The first generation created from the cross was sterile, so the cell division inhibitor colchicine was used to double the number of chromosomes in the cell. Cells with an uneven number of chromosomes are sterile.

Failure to produce a hybrid may be due to pre- or post-fertilization incompatibility. If fertilization is possible between two species or genera, the hybrid embryo may abort before maturation. If this does occur the embryo resulting from an interspecific or intergeneric cross can sometimes be rescued and cultured to produce a whole plant. Such a method is referred to as *Embryo Rescue*. This technique has been used to produce new rice for Africa, an interspecific cross of Asian rice *(Oryza sativa)* and African rice *(Oryza glaberrima)*. Hybrids may also be produced by a technique called protoplast fusion. In this case protoplasts are fused, usually in an electric field. Viable recombinants can be regenerated in culture.

Chemical mutagens like EMS and DMSO, radiation and transposons are used to generate mutants with desirable traits to be bred with other cultivars.

Classical plant breeders also generate genetic diversity within a species by exploiting a process called somaclonal variation, which occurs in plants produced from tissue culture, particularly plants derived from callus. Induced polyploidy, and the addition or removal of chromosomes using a technique called chromosome engineering may also be used.

When a desirable trait has been bred into a species, a number of crosses to the favoured parent are made to make the new plant as similar as the parent as possible. Returning to the example of the mildew resistant pea being crossed with a high-yielding but susceptible pea, to make the mildew resistant progeny of the cross most like the high-yielding parent, the progeny will be crossed back to that parent for several generations. This process removes most of the genetic contribution of the mildew resistant parent. Classical breeding is therefore a cyclical process.

It should be noted that with classical breeding techniques, the breeder does not know exactly what genes have been introduced to the new cultivars. Some scientists therefore argue that plants produced by classical breeding methods should undergo the same safety testing regime as genetically modified plants. There have been instances where plants bred using classical techniques have been unsuitable for human consumption, for example the poison solanine was accidentally re-introduced into varieties of potato though plant breeding.

Exploiting heterogeneity and crop evolution in farmers' fields are outside the scope of most plant breeding research. One exceptional experiment, however, has shed some scientific light on the issue. It was started at the University of California (UC) in 1928. Composite cross populations of barley were produced, some of which were extremely diverse in origin of sources. These populations were exposed to continuous natural selection in current modern farming environments and became the subject of studies during the career span of several generations of UC professors.

It appears that after low yields in initial years, the composite cross populations gradually improved in performance and eventually became quite good yielders, with excellent yield stability and disease resistance. These results inspired Suneson to propose an evolutionary plant breeding method. After assessment of later generations of the same material, Soliman and Allard concluded that such evolutionary breeding "is unwarranted" if yield potential is the major goal. However, if disease resistance and yield stability are two major objectives, "the composite cross approach is an efficient method". This amounts to saying that a major part of world agriculture, many high-input systems included, could be well served by this approach.

In short, this means constructing a body of broadly diversified germplasm and exposing it to natural selection in areas of contemplated use. For those who are familiar with traditional farmers' breeding, this may sound like reinventing the wheel. In fact it is an improvement of the old wheel of plant breeding.The first step, constructing a body of of broadly diversified germplasm, is not all that straightforward. Science has access to world collections and information sources that are unavailable to farmers. A research institute can chose relevant germplasm and make composite cross populations with an evolutionary potential, most probably far beyond that of locally available varieties.The immediate outcome, the early generation composite cross population, will be unadapted everywhere and is likely to yield poorly. With time, however, recombinations and natural sorting will improve the adaptation, and, according to the Californian experience, narrow the gap with commercial varieties. The long term outcome could be populations that outperform commercial varieties in disease resistance and yield stability and that may be used as a source of artificial selection for high yield.

The disease resistance appears to have evolved through the building up of polygenic complexes. Therefore, it provides a durable resistance as opposed to the monogenic and, therefore, mostly non-durable resistance usually bred into commercial pure line varieties. The stability, at least to some degree, depends on the buffering effect of crop heterogeneity. The Californian experiment shows that when the population is propagated in isolation for a very long time, diversity will start declining, resulting eventually in reduced stability.

These experimental findings into the context of current development needs, a few conclusions can be drawn. We need breeding populations with a very high evolutionary potential, and these populations must be exposed to the stress conditions of, or similar to, current farm environments. Furthermore, a certain level of diversity within populations must be maintained in order to sustain evolutionary potential and yield stability.

An Age-old Tradition

Modern plant breeding is a sophisticated, high investment business, but its origins stretch back thousands of years to primitive farmers who selected the best plants in one year to provide seed for their next crop. This selective breeding was the first human refinement of natural plant evolution. Recent scientific and technological developments have allowed a greater rate of improvement. It was Gregor Mendel who, in the mid- 19th century, first provided a scientific explanation of genetic inheritance. His conclusions on the relationship between inherited characteristics in the offspring and the genetic makeup of the parents were the theoretical basis for classical plant breeding. Mendel's work went largely unrecognized in his own lifetime, and it was not until the early 20th century that it was rediscovered to become the basis of modern scientific plant breeding.

A Modern Industry

Until the early 1960s, plant breeding in Britain was largely confined to publicly funded research. This situation changed dramatically in the mid-1960s, with the passing into UK law of the 1964 Plant Varieties and Seeds Act. This legislation introduced a system of royalty payments on individual plant varieties, known as Plant Breeders' Rights, and triggered a rapid expansion of plant breeding as a commercial enterprise in its own right. Today, much of the basic research into crop science is still conducted by public sector research organisations, but the majority of commercial plant breeding takes place within the private sector. Some

60 plant breeding companies, based in the UK, are active across the entire spectrum of plant species from the major arable crops through to ornamental garden shrubs and flowers. In total, the plant breeding sector employs around 5,000 people, and supports a further 5,000 jobs in seed production and distribution. Plant breeding remains a vital industry to keep Britain competitive on world markets. The need for new varieties, adapted to our unique growing conditions, is never ending, driven by the challenges of new disease pressures, changing market requirements and shifts in agricultural and environmental policies.

- *Maize*: Maize crops used for grain and animal fodder are derived from wild races originating in Central America
- *Sugar beet*: Modern varieties of sugar beet have been developed from wild ancestors native to Central Europe.
- *Potatoes*: Wild ancestors of the modern potato still grow in parts of South America.
- *Wheat and Barley*: Small grain cereals, the mainstay of UK crop production, are derived from wild grasses of the Middle East.
- *Oilseed Rape*: Like many crops within the brassica family, oilseed rape has its origins in wild species native to China.

Participatory Plant Breeding

Certain trends have made the world ripe for adoption of participatory plant breeding methods.

- The shielding of crops from environmental stress in high-input systems is facing increasing economic and ecological problems. Scientists are changing their attitudes, and a new paradigm is being formulated. Instead of modifying the environment to suit the requirement of high yielding varieties, the varieties need to be modified to suit the environment.
- The claim that modern varieties can be made broadly adapted and be superior across most farming environments within an ecogeographic region is being challenged.
- If relevant diversity exists in a locale, the combined action of natural and artificial selection within a local environment may be an efficient

breeding method. Experiments show that this may work also in a fertilizer-intensive system.

- In recent years farmer groups working with local seeds have been organized all over the world. They are not primarily conservers of old seeds. They want their seeds to be improved, in an evolutionary, slow and steady way, and under their own control.
- Finally, participatory methods have been developed in order to facilitate the involvement of farmers together with scientists as active and equal partners in research to generate relevant farm technology. Such methods can be applied also to plant breeding.

Commercial seeds often diffuse into areas where the traditional seed supply system is still predominant. Farmers try them with an open mind and adopt or reject them according to their own criteria. If grown and multiplied in the villages, diversity will start to appear within them and local reselection will be possible. In that way commercial varieties eventually might become like landraces. It is also commonly observed that farmers change and exchange seeds. A traditional farming system rarely functions as an environment for the static preservation of old landraces.

Often practices range from neglect to very simple mass selection, but with a few scattered individuals who devote an exceptional amount of effort to the maintenance or improvement of seed quality. These exceptional persons, very often women, may be the source of good seeds for others in the community. Once such a community is organized for seed management and improvement, it becomes possible for it to establish links to scientific institutions.

Community resources for participatory approaches to seed management and breeding are not limited to indigenous culture and traditional practices. The educational status and experiences of modern farmers may also be turned into a resource for community action. In the Philippine group, a number of the members were high school graduates and a few had a university degree. And moreover, most of them had a couple of decades' experience with modern input-intensive farming. Seed activities opened their minds towards the traditional societies, towards themselves and towards the modern world. The traditional societies supplied them with their seeds, and through the seeds, they learned to appreciate the values and achievements of these societies. They discovered their own potential, and saw that the outside world could bring more than technology packets: it could bring knowledge and ideas to be exploited and further developed by themselves.

TECHNIQUE OF CROSS-BREEDING

Here is hardly any plant-breeding project which does not involve cross-breeding. In most cases it is the object to produce variability, in order to obtain the material for selection to work upon. The quickest way to obtain variability is cross-breeding.We should distinguish between the work of the plant breeder and that of the scientific worker. It is true that the great majority of plant-breeding problems are tackled at the larger institutes by genetically trained specialists, and yet the distinction should be made. In scientific experiments only the most painstaking precautions are good enough. We must be absolutely certain about the paternity in our hybrid plants, and for this reason both the collection of the desired pollen and the elimination of the pollen from the mother-plant must be done with the minutest precautions.

If our object is a wholly practical one, we may often relax such precautions to a certain extent. For instance, in some plants castration—the elimination of stamens—is not only extremely difficult, but the manipulation brings a risk of seeing the flower wither and die. In such plants (soybeans, oats) castration can often be dispensed with, when we choose the parents in such a way that we can recognize the hybrids from inbred plants.

In every plant-family our first step must be to study the biology of the flower, in so far as this has not been done by others. We must know at what stage of growth of the flower- bud the anthers dehisce, so as to hit upon the safe time at which we must perform the operation of removing them. We must also know when the stigma is receptive, to find the moment at which to apply the pollen. In most plants castration and the application of pollen can be done at the same time. This is the case in rice, in most of the cereals, in tobacco, Ipomoea, cocoa, peas, beans and a host of other plants. In some plants with large flowers the stamens are attached to the corolla, and it is possible to cut off the corolla with the stamens in one operation. This can some- times be done with the fingernail (apple, raspberry, straw- berry) or with a scalpel.

When a scalpel is used, it is sometimes advantageous to cover the blade almost up to the tip with adhesive tape, to avoid cutting down too deep. Where this is impossible, the stamens have to be extracted one by one. A set of fine pincers with differently curved tips should be in the possession of every worker. Very useful fine instruments can often be made to fit particular occasions. A very fine hook, made out of a bent pin, and mounted into a needle-holder can sometimes be used to great advantage, especially in castrating such minute flowers as in alfalfa or clovers.

For fine work a watch-maker's eyeglass, or even a binocular dissecting glass, may come in useful. On the whole, dispensing with castration in plants with minute flowers, and to use the method of straight cross-breeding, leaving the finding of hybrids till next season.

In some plants male and female flowers are separate. In such plants no castration is necessary; all we have to do is to protect the flowers from unwanted pollen. In cucurbits, melons, squashes, cucumbers, we found it quite easy to close unopened flower-buds, by means of small pieces of copper, or preferably lead-wire. Of course both male and female buds should be so protected. Lead-wire has the advantage of being very soft, and it can be used many times in one season.Especially in wind-pollinated plants, care should be taken to protect the pollen used against contamination with unknown pollen. For this purpose the flowers from which pollen must be collected should be bagged. In some plants pollen is so abundant that we can use a separate anther for every flower to be pollinated, and we can hold this with fine forceps. Where the pollen is dry, it can be shaken out of the anthers into some small receptacle, for which a small gelatine capsule is just the thing. Such capsules can be closed, and the pollen contained in them can often be kept in very good condition for days, if the labelled capsules are kept cool in a well-closed bottle over calcium.

In maize both the male and female inflorescence should be bagged early. When the silks protrude they can be trimmed off short, and the pollen shaken over them from a tassel. If it is wished to obtain a cob full of kernels, it is of some advantage to insert a flowering tassel into the bag that covers the cob, and we can prolong its life and usefulness by standing this tassel in a small bottle of water hooked to the stalk.

In certain plants the operation of castration and fertilization can be performed on cut-off stalks that have been taken into the laboratory. In sugar-cane, in millet and sorghum, and even in wheat and barley, such panicles and ears, when stood in water, will ripen their seeds well, especially if only a few flowers are left on them for this purpose. Paper bags should be of ample size, and it is necessary to leave them on no longer than is absolutely essential for full protection during the few days in which the stigma is receptive. In certain cases (coffee, cocoa) a finely woven tartan may have advantages over paper; it is softer, and can be arranged around awkwardly placed flowers better than paper. Of course it is useful in certain cases to give the ripening fruit or seeds some protection against loss or damage (stone-fruit, potato, tobacco, (Enothera), and paper bags may be necessary at that stage.

For keeping small quantities of seeds, small paper or preferably cellophane envelopes can be used. Gelatin druggist capsules, however, are preferable. The labels should be en- closed with the seeds.

LABELLING, NUMBERING AND GADGETS

Every plant-breeder and every geneticist has his special methods of keeping his records. To be able to do this he must have suitable methods of numbering or labelling his plants or seeds. Any system of labelling or marking must be legible and in- destructible by weather or insects. For merely marking, as an aid to tracing a special plant or part in a field, a coloured thread may be quite sufficient. If such a thread is used in addition to a number or a label, it helps to distinguish that label. Paper labels, even if made from very good parchment paper, tend to become discoloured. To depend on such labels only, let us say in marking pea-pods, is courting trouble, for they closely resemble dried leaves or empty pods. The well-known Vilmorin red woollen thread is of very great value; it may even serve the double purpose of protecting the cross-fertilized ear of wheat or barley and of making it easy to find it again. To mark seed capsules, fruits and such things, we found that lead-wire was much better than string. We can string a glass or china bead on a small piece of very soft lead-wire, and if we need a few dozen distinguishing marks, combinations of coloured beads will be excellent. We used this method in marking cross-fertilized squash blossoms and melon blossoms. If we make a loop of sufficient width, we can squeeze one end of this against the fruit-stalk, and it will accompany the fruit until harvesting time, or even beyond.

A very good label that will never lose its legibility is a poultry leg-band with a perforated number. These are made out of thin aluminium, and can be opened and adjusted before closing. Such labels are just the thing to mark plants that must be taken up and wintered in a cellar, like dahlias.

Most seedsmen use stout painted stakes for carrying the name and the number of rows and individual plants, the name facing the row it marks. To obviate mistakes it is safer to have duplicate stakes, or to have the name and number of each lot written on a separate stake* To save time and worry, good use can be made of poultry leg-bands or of the newer coloured plastic numbered wing- shields. These are numbered on both faces, and can be bought in consecutively numbered hundreds or thousands.

They can be hung from wire stakes with eyelets, and they accompany such plants as cereals or beans when they are bundled for harvesting. In some countries one has to guard against theft and mischief. A European blackbird pull out dozens of numbered yellow wooden labels and scatter them all over the place. A system of book-keeping which allows one to find the missing numbers from the book is indispensable if we want to avoid mistakes.

Numerous methods have been invented for protecting flowers and branches against insects or wind-borne pollen. Paper bags come first; they must be strong, rainproof and light. Some cotton-wool may come in very useful in making the connection of bag and stalk pollen-proof. Glassine bags and lengths of seamless sausage casing are being widely used in the United States. They allow one to see the inside of the bag, and so may save much time and error.

In some sympetalic plants flower-buds may be closed hermetically by the application of a length of very soft wire (lead- wire) wrapped round the bud, and used again after fertilization. We have used this method with squashes, melons and Ipomoea, The American plant-breeding journals and publications abound with methods for castrating and cross-breeding plants. Removing the pollen from the stigmas by washing it off or killing it by heat are methods but they sound excellent. Gelatin capsules, made in two pieces, of which one fits over the other, are handy receptacles for keeping very small quantities of seeds or of pollen, for slightly larger quantities sachets for internal use may be used;

they are cheap, and can be written upon, but if we close them we can never use them again. Paper bags must be used a good deal in storing seeds and similar objects (bulbils, bulbs, corms). They can be closed in different ways. The simplest system is the method of the French seed firms, in which each bag is closed by three folds, in such a way that it will never open spontaneously, even if submitted to considerable shaking.

The gadgets used in castration and cross-pollination are very numerous, and must be adapted anew to every new plant. Of course it is necessary to kill adherent pollen with alcohol, and extra scalpels, forceps or scissors are of great help. In working with very minute flowers a good binocular microscope is invaluable.

Avoid castrating minute flowers as much as possible. Everyone doing this sort of work occasionally regrets he was born with only two hands and without a prehensile tail. A ring to which has been soldered a minute cup that holds a small gelatin capsule with pollen, and that can be worn on the index-finger of one hand, is almost as useful. This is another American invention.

Hard seeds can be treated in different ways to make them grow. Some seeds, like those of canna or musa, can be scratched with a file, others may be rubbed on stone. Enothera seeds can be soaked in water; some seeds will only take up water rapidly when we exhaust the air above the water in which we keep them and then readmit the air. Seeds of apricots, cherries and plums can be extracted after carefully cracking the stone. The extracted embryos (and the same is true for embryos of apples, pears and citrus) must be protected against bacteria and mould. Thymol solutions, even weak ones, will do this without injuring the seedlings.

SPEEDING-UP PLANT-BREEDING WORK

Plant-breeding work involving crosses generally takes a great number of generations before the superior new plant has been obtained. In the routine work of a large plant-breeding station so many breeding projects arc under way simultaneously that it matters little how much time elapses between the initial cross and the final purification of a promising commercial variety. Every year there are many projects being started, several are under way, and a few arc in their last stages.

It often pays to cut down the time required for work of this kind. One of the most promising means of doing this is to grow more than one generation of plants in one year. This usually means that one of those generations of plants is grown in a season in which the plant is not ordinarily grown, and that for this reason a correct evaluation of the quality of the individual plants is out of the question.

In Europe or in America growing an extra generation on the spot means growing plants in the greenhouse. To grow cereals or beans in winter in greenhouse conditions not only requires heat; we must also have the means of regulating the length of day. The simplest method is to use large electric lights to furnish both the heat and the light, to lengthen the day by those means, and to let the temperature go down during the dark part of the night. Under those conditions cereals and beans will flower and set seed very well in the winter months,

The electricity required per square yard is roughly that used by lamps that require a kilowatt. It is clear that under such conditions the winter generation will not show the normal qualities of the plants in that group which are grown in summer-time in the field, but there are many cases in which this is not of much importance. If we want to proceed by the method of making some hybrids and then growing a second and third inbred generation from this stock, we can do our crossing in the field, grow our hybrids in the winter, and then grow another crop of second-generation plants in normal field conditions. When we are using the method of mating back hybrid stock to some pure kind of plants we want to improve it matters but little if we cannot very well select the best plants during some of the generations involved. It is evident that we must grow a generation under normal field conditions whenever we want to make our selections.

In recent years a few seed-firms have hit upon the perfect scheme of growing two generations of annual plants. This does not involve any greenhouse generations. They work in two different localities with very similar climate, but so chosen that one of those localities is situated in the Southern Hemisphere.

There are regions of Chili where Californian firms can find conditions of plant-growth that are identical with conditions at home, with the exception that the growing season comes during the Californian winter. In those circumstances it is quite possible to grow two absolutely normal generations during one year. Our present greatly improved methods of communication make such schemes perfectly feasible.

With many tropical plants one generation yearly is grown of plants that need special conditions of either a wet or a dry season. Here it is easy enough to interpolate an additional generation, by looking for a suitable spot, or by providing the necessary irrigation. Sometimes it takes a very long time to bring the plants to a condition where we can evaluate them. A good example is that of the fruit trees. It takes many years for an apple seedling or a seed-grown cherry to come into bearing. Here we can save a great deal of time by budding or top-grafting parts of the seedlings upon mature trees in full bearing. By this method we may save more than half the number of years normally required for bringing the seedlings to maturity.

In this connection we might also treat of those cases where certain seeds take a very long time to germinate. Storing the seed in suitable conditions of moisture, and especially of temperature, will speed-up germination in such plants as gooseberries and celery. In stone-fruits it has been found possible to crack open the stone, to extract the kernel and grow the seedlings under aseptic or even under antiseptic conditions (thymol solution).

SHOWS AND SHOWING

The exhibiting of plants and seed at the shows is a very curious thing. It pays extremely well for a seed-firm to employ some gardeners who know just exactly when and' how to get enough plants full of luscious beans or covered with flowers on the exact day for each show. Florists and lovers of gardening see the quality of such plants at the shows, and many people who admire them there try in their turn to win the coveted prizes at their local shows the next year. This show game is certainly very nice and pleasant; it gives the gardeners something to strive after; it may even help to keep some of the lads in the country.

On the other hand, there is very, very little connection between the qualities judged at the shows and actual value of the material shown. The few selected mangolds, or the mammoth pumpkins, or the bundle of wheat- ears, or even of the maize-ear contests of the American agricultural shows.

All along the line, in horticulture as well as in agriculture, the selected samples shown give no indication of the value of the seed or plant stock the producers have for sale. Of course we can see that those few mangolds have the ideal shape and size, but the seed-firm must give us some guarantee about the percentage of mangolds that will grow to this size and shape from the seed they sell, just as the owner of a prize-winning bull must give some guarantee of this animal's ability to give profitable daughters, in addition to the blue ribbon awarded to the beautifully coloured and shaped prize-winner.

In agriculture this is beginning to be so well realised that the classes for beautiful corn cobs or for the largest mangolds are becoming very rare indeed, even if we still see stock-judging in dairy cows and similar farces. In the horticultural shows much interest is still shown in regard to flowers and vegetables.

It seems evident that when we strive for beauty, beauty contests for roses and begonias, for asparagus and delphiniums and dahlias are indicated. The show is a shop-window. The seller must show if his competitors show, if only to keep his name before the public. To the public the show is only just a pleasant occasion for a half-day outing. To the buyer the show is much nicer to see than a description in a catalogue.

Long lists of novelties are bought at the shows by people with an experimental turn of mind. We see the representatives at the shows kept busy noting down long lists of orders. Showing certainly pays the seller. A very much more sensible system of showing a great many varieties is that of the demonstration gardens and experimental stations, both stations in which the agricultural authorities try to compare promising new and good old varieties with the aid of objective standards, and such gardens as show-plots arranged by rose-lovers or by societies of growers of gladioli or dahlias. Here the varieties can be judged throughout the growing season in favourable conditions.

THE ANNUAL SELF-FERTILIZING CROPS

The number of plant species now being cultivated by man is really enormous, while only relatively few animal species are kept under domestication. It is possible for the author of a book on animal breeding to take those species almost one by one, and to give the specialist breeders a few hints over and above those which are contained in the general book. This would be quite out of the question in a book on plant breeding. One way out of the difficulty would be to give one example of each of the three groups—self-fertilized plants, cross-fertilizers and plants that are usually propagated by vegetative means. These plants in two sections— grains and legumes—giving some hints to the breeders of wheat, barley, oats, sorghum, rice, etc., and going into some details about the breeding of beans, soybeans, cow-peas, ground-nuts, peas, etc., separately. But, after all, the self-fertilizing annuals have so much in common from the point of view of plant-breeding methods.

Self-fertilization causes isolation. Each plant, as far as reproduction is concerned, is as self-contained and isolated in a crowded field as it would be if grown by itself in a greenhouse in a city. The inbreeding caused by self-fertilization has several results. It makes all heritable variability disappear in every line, it makes every group consist of parallel lines, in which each plant descends from only one parent plant, so that the usual complicated network of ancestry is simplified into a system of isolated parallel lines. Where inherited variability exists, such as happens after a deliberate artificial cross or when accidental cross-fertilization occurs, it will automatically disappear.

That is to say, it will disappear in so far as each separate line is concerned, for Mendelian segregation will make the lines, descending from any hybrid heterozygous for a great many genes, diverse in their hereditary make-up. The result is that a field of self-fertilizing annuals that has not been grown from just one pure line by a plant breeder always consists of a mixture of a number of genetically different lines. Even if accidental cross-fertilization or irregularities in crossing-over (mutation) are very rare, they do occur, and cause this state of things. This genetic variability, however, is wholly different from what we find in other plants and animals. We are dealing with mixtures of pure lines, and almost every plant we take from any field is pure, homozygous for all its genes, and will, if its seeds are sown, give a very homogeneous descendance. From all this it follows that enormous progress can often be made in all self-fertilized plants by simply picking out a great number of promising-looking individual plants and comparing their descendance.

Many valuable widely grown wheats, tobaccos, peanuts and beans have been just found. They "just growed ", like Topsy. The first step in plant-breeding in the self-fertilizing plants must always be to look for those ready-made pure lines. The next is to find whether a mixture of two of the very best pure lines can be discovered which will give us still higher production (10 per cent, higher production from a mixture of two strains as compared with that of the best one is quite common).

We have then to produce some more genetic variability by deliberate cross-breeding. This generally means that we try to combine the good qualities of two different strains in one new one. Here there are two wholly different schools of thought, and two wholly different methods. One of them is to analyse second-generation plants, and to find those individuals that do combine, let us say, the disease resistance of line A with the large quantity of big seeds of line B. This is a method in- vented by the experimental geneticists who are trying to combine practical plant breeding with purely scientific gene- analysis. As a theoretical geneticist, interested in genes and their action.

It is great fun playing with a hundred different hybrid lots in one species, growing a few dozen second-generation plants in each lot and tabulating the results. But this is neither sound genetics nor sound plant breeding. We are dealing with so many unanalysed and unanalysable genes which all affect production and quality at the same time, that the chance of finding something really worth while in a few hundred plants is simply not good enough.

The other method is based upon the realization that hundreds of genes are commonly involved, and upon the fact that purification—reduction of the potential variability—is rapid and automatic in this material. It consists of growing the greatest possible number of seeds from first-generation hybrid plants. We can then proceed in two different ways. One is an analytical method, which means that we are seeking for the very best-looking plants (for instance, by weighing their seeds), and progeny testing those plants by sowing a large number of seeds from each, in order to find the most profitable families. The other method is what Baur and Nilsson-Ehle called a "Ramsch" method.

This consists simply of growing the descendance of a hybrid in a vast mixture, refraining from any analysis, and continuing this hybrid lot as a miscellaneous collection for five or six generations, sowing as much of the mixed seed of each harvest as we have room for. If we do this, natural selection will weed out most of the unproductive and undesirable types, such as those plants tl at ripen after harvesting time, all weaklings and all plants susceptible to diseases and pests. After five or six years that lot will consist of a mixture of pure lines, every line different from every other, with the successful lines represented by more individuals than the unsuccessful ones. After this we will treat the mixture just as if we were dealing with any other mixture, growing separate rows of beds each from one likely-looking plant. This is a very sound method, which saves a great deal of un- necessary labour and time. It is extremely difficult to find just how much better this method is as compared with that of analysis in every generation, starting from the second generation. An analytical method could be made to work very well, provided we did all we could to make the number of plants of the second and third generations exceedingly large (in the cereals we can get almost as many seeds as in tobacco by splitting up the hybrid plant repeatedly before it starts to grow any stalks).

In actual practice there are only a few crosses that really produce anything worth while, and the general practice seems to be to make a great many attempts and grow a large number of first-generation hybrids. This renders it imperative to cut down on the number of plants per experiment. It is a vicious circle. This is all very wrong and wasteful. One cross every year and work with it, both analytically and by means of a "Ramsch ", doing the final selection after six years for one lot in every season, than waste my time castrating flowers and producing numerous hybrids that would have no really good chance of showing what they could do.

In all the self-fertilizing annuals we must be continually on the look-out for qualities that may be of special merit. Mostly of adaptations to special conditions. In wheaa on irrigated ground, resistance to drowning may be as important as disease resistance or quality of the grain. Conversely, rice could be grown much more extensively if we would take the trouble to do some plant-breeding work on dry-land rices, which abound in the mountain villages of all the tropical Asian islands. In ground-nuts we should look for lines that are specially adapted to mechanical harvesting, as well as for lines adapted to the production of a first-quality product in the hands of tens of thousands of small cultivators. In all the plants of this group it is possible not only to improve the yield and the quality, but also to extend the area in which profitable production is possible. Tobacco, rice, soybeans and cow-peas are just as important in this respect as wheat, barley and oats.

Plant breeding with self-fertilizing plants is comparatively easy. It is even possible for the real amateur plant breeder to do some extremely interesting and useful work with some of the less well-known plants of this group—let us say with beans or cow-peas. Such work needs room, but very little special technical knowledge or talent for higher arithmetic. Frequently a few plants will stand out from an otherwise quite even field of barley, beans or sorghum. Such "high birds "—climbing plants in bush beans, high, scmiawned plants in wheat—are often accidental hybrids. If we sow a few of their seeds, we can see immediately whether they are hybrids or admixtures, for hybrids will have very variable offspring. If then we sow all the seeds from such hybrids, we can start a series of experiments that may easily give us something worth while, even when the father of the hybrid is unknown.

Scope of plant breeding

Improving Crop Performance

The dramatic gains in agricultural productivity seen in the second half of the last century are often linked uniquely to increased mechanisation and the widespread introduction of fertilizers and agrochemicals.This 'Green Revolution' would not have been possible without the huge genetic improvements made to the crops. Plant breeding has contributed around half of the threefold increase in UK wheat yields recorded from 1947 to 1992 and yields have continued to improve since then. Plant breeding can directly improve the performance of crops in different ways.

Productive Yield

Developing crop varieties, which convert more of their biomass into productive yield, is the single biggest contributor to improved crop output. The introduction of shorter-strawed cereals is a striking example of how this has been achieved, by transforming more of the crop's productive energy into valuable grain.

Physical Characteristics

Changing a crop's physical structure can also contribute to increased yields. For example, the development of semi-leafless varieties of field peas helped to boost intrinsic yield. It also stimulated the growth of stabilising tendrils, which significantly improved the crop's standing ability so reducing crop losses at harvest.

Disease Resistance

Genetic resistance to disease enables crops to realise their yield potential– it can also mean reduced use of agrochemicals. Plant breeding has significantly improved the genetic resistance of crops against the threat of viral and fungal infection. Key examples include resistance to blight in potatoes, rhizomania in sugar beet, and barley yellow mosaic virus in cereals. The challenge of breeding resistant varieties is constant because new strains of disease develop naturally.

Timing of Maturity

Plant breeding technology has brought major improvements in the uniformity with which crops ripen ready for harvest. This not only reduces potential crop losses at harvest (as in the case of pod shatter in oilseed rape), but has also improved growers' ability to mechanise harvesting operations. In the field vegetable sector, for example, the labour intensive (and unpleasant!) task of picking crops such as cauliflower, broccoli and Brussels sprouts has been transformed by the development of improved varieties.

Other Agronomic Factors

By improving crops' ability to cope with a range of other agronomic pressures, advances in plant breeding continue to underpin progress in agricultural productivity.

They include:

- Genetic resistance to pests, such as nematode resistance in potatoes
- Shorter crop life-cycle, to expand a crop's geographical growing areas
- Stress tolerance, such as frostresistance in field vegetables, to extend the seasonal availability of home-grown fresh produce.

Improving Crop Quality

Alongside progress in keeping Britain's farmers competitive through improved productivity, developments in plant breeding have also brought significant gains in crop quality. Gone are the days when Europe encouraged farmers to concentrate solely on maximising output, assured that their crops would find support under the Common Agricultural Policy.

Today, as production-based support and barriers to international trade continue to be reduced, Britain's growers must compete with the best in the world, on quality and cost of production. Consumer expectations of food are also more exacting. This is reflected in tighter quality specifications along the food chain. Plant breeders have responded with a continuous stream of new varieties, tailored to the needs of specific end-markets.

Cereals

Britain's historical dependence on North American imports of wheat for bread making has been reversed over the past 40 years thanks to improvements in baking technology and the development of high protein, hard milling varieties of wheat.

Until the mid-1960s, imported wheat typically accounted for up to 65% of our bread making requirements. Today as much as 90% of our bread is produced from homegrown varieties. In barley, improvements in malting quality have underpinned a fourfold increase in the volume of beer brewed per tonne, from around 2,000 liters in 1950 to nearly 8,000 liters today.

The quality of malting barley varieties grown in Britain is internationally recognized, with export markets from the near Continent to the Far East.

Oilseed Rape

Oilseed rape has become a major homegrown source of vegetable oil and animal feed within the past 30 years. The rapid expansion of this important break crop from 50,000 hectares in the mid-1970s to around 500,000 hectares today can be linked to two major breakthroughs in plant breeding. Varieties for food use were first developed in 1970. These contain reduced levels of erucic acid in the oil, making it more suitable for human consumption. This was followed in the late 1980s by 'doublelow' varieties, which also offered reduced levels of glucosinolates in the residual meal to improve quality for animal feed.

Potatoes

In the potato sector, plant breeders have responded to increasing consumer demands for quality and choice. Fresh produce counters now boast a bigger range than ever, from traditional main-crop varieties through to baby new, baking and salad potatoes. Breeding programmes have also targeted processing and catering markets, with specific varieties used to produce crisps, instant mash and chips.

Sugar Beet

Higher root yield and increased sugar content have combined to double sugar production per hectare in the past 50 years. The introduction of monogerm seed, allowing seeds to be planted individually rather than in clusters, ensures genetic improvements are fully realised in the field.

Pulses

As interest in home-grown sources of protein has increased, breeders have developed varieties of peas and beans for specialist markets–freezing and canning for human consumption, flaking for pet food, and tannin-free varieties for animal feed.

Creating a New Variety

The creation of each new variety is a complex, costly and skilled operation. It is also painstakingly slow–today's breeding programmes are already looking ten years ahead to the needs of farmers, consumers and the environment at the end of the next decade and beyond. Techniques vary between crop species, but the scientific principles of plant breeding remain unchanged from Mendel's first discovery that selected parent plants can be cross-pollinated to combine desired characteristics. Genes–units of hereditary material that are transferred from one generation to the next, determine plant characteristics. Since each plant contains many thousands of genes, and the breeder is seeking to combine a range of traits in one plant, such as high yield, quality and resistance to disease, developing a successful variety is an extremely lengthy process–up to 12 years in the case of cereals, even longer for potatoes. Plant breeding has been compared to playing a fruit machine–not with three reels, but several hundred. The skill of the plant breeder lies in improving his chances of hitting the jackpot by combining all the desired characteristics in the same variety.

6

Micropropagation

In vitro plantlets, which are free of pathogens, are used as initial material for potato and sweetpotato seed programmes. The methods used in these micropropagation programmes mainly depend on their production volume and the available infrastructure. In the case of potato micropropagation, the basic methods have been already described. They have been verified in many institutions and they are based on the rapid growth of individual node cuttings, or stems with multiple node cuttings. Afterwards, the basic micropropagation methods are described.

NODE MICROPROPAGATION

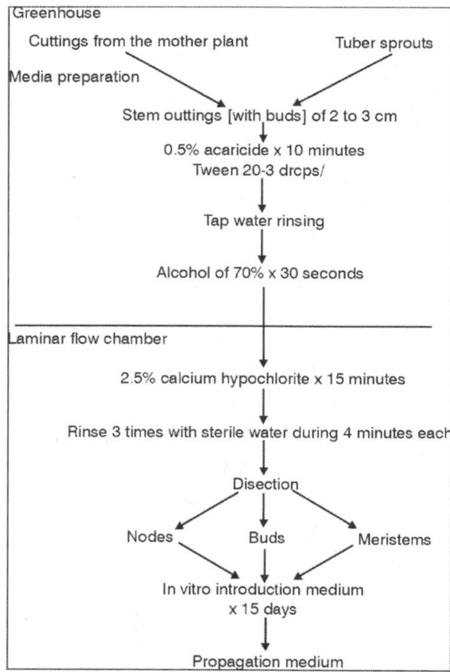

Fig. Procedure for in vitro Introducction

This method is based on the principle that the node of an in vitro plantlet placed in an appropriate culture medium will induce the development of the axillary bud, resulting in a new in vitro plantlet. This type of propagation promotes the development of a pre-existent morphological structure. The nutritional and hormonal condition of the medium breaks the dormancy of the axillary bud and promotes its rapid development.

Callus formation and plant regeneration must be avoided because they tend to affect the genetic stability of the genotype. Under room-controlled conditions micropropagation is fast. Each node planted in a propagation medium will produce a plantlet which will occupy the full length of the test tube, after approximately four weeks for potato, and six weeks for sweetpotato. The resultant in vitro plantlets may be transplanted to in vitro conditions in small pots in the greenhouse.

MICROPROPAGATION BY NODE CUTTINGS IN A LIQUID MEDIUM

This technique is applied both with potato and sweetpotato to produce a large number of nodes rapidly. Stem cuttings with 5 to 8 nodes are prepared by removing both the apex and the root of the in vitro plant to be propagated. The stems are placed in the corresponding propagation liquid medium. It is also possible to use isolated nodes: the nodes will sprout and new plantlets will develop over a period of 3 to 4 weeks.

Micropropagation Procedure

- Sterilize petri dishes (placed in paper bags or comets) and prepare the laminar flow chamber by disinfecting the internal surfaces with alcohol.

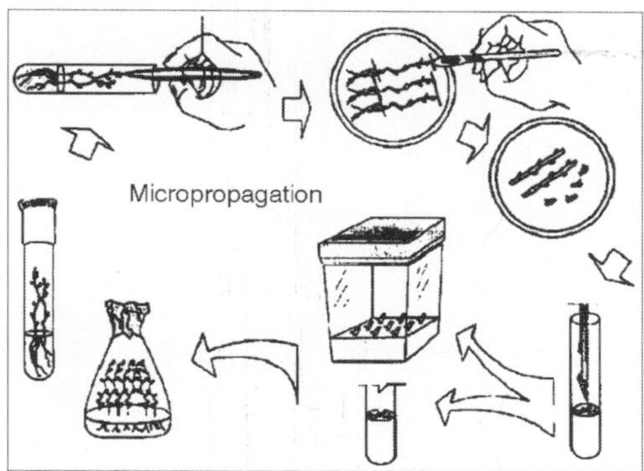

Fig. Potato Micropropagation Process Scheme

Sterilize the tools with an instrument sterilizer and place them on a sterile dish.

- Open the tube, take off the plantlet and place it on a petri dish with the help of forceps.

- Remove the leaves and cut the nodes.
- Open a tube containing fresh sterile medium and place a node inside, trying to plunge it slightly into the medium with the bud up. Close the tube.
- Seal the tube with a gas-permeable plastic tape (parafilm or saram wrap) and label it correctly. It is recommended to place two explants in 16 × 125 mm tubes, three in 18 × 150 mm tubes, five in 25 × 150 mm tubes, and 20-30 in magenta vessels.

COMMON PROBLEMS IN MICROPROPAGATION

Some problems may appear in tissue culture according to the crop or variety. To solve them, it is necessary to apply one or several preventing/solving methods such as:

Phenolization

The explants frequently become brown or blackish shortly after isolation. When this occurs, growth is inhibited and the tissue generally dies. The young tissues are less susceptible to darkening than the more mature ones.

Prevention

The tissue darkening– mainly that of the recently isolated explants and that of the medium–may be generally prevented by:

- Removing the phenolic compounds produced by dispersion. Absorption by means of activated carbon. Absorption by polyvinilpyrolidone (PVP).
- Modifying the redox potential. Reducing agents: ascorbic acid, citric acid, L-cisteine HCL, ditriotreitol, glutation and mercaptoethanol Less availability of oxygen: stationary liquid media.
- Inactivating the phenolase enzymes. Chelating agents: NaFeEDTA, EDTA, diethyldithiocarbamate, dimethyl-dithiocarbamate
- Reducing the phenolasic activity and the availability of substrate. Low pH Darkness

Absence of Rooting

The explants can naturally form roots during propagation, without an additional rooting stage, as with the potato. However, some wild potato species may show root production deficiency. Rooting may be induced by incorporating auxins, such as IAA, NAA, and IBA, or activated carbon to the culture medium.

POTATO IN VITRO TUBERIZATION

Most of the potato microtubers are used in seed programmes in Europe, where large amounts (hundred of thousands) of pre-basic seed of a few varieties are produced. By means of this technique, microtubers are produced and stored, and it is possible to store thousands of them in a small area (in humid containers at 4°C) for long periods of time. The microtuber induction is produced through a stress effect by the CCC (chlorocholine chloride), BAP (B-benzylaminopurine) and Sucrose, which under darkness will produce from 3 to 4 microtubers per plant, according to the variety.

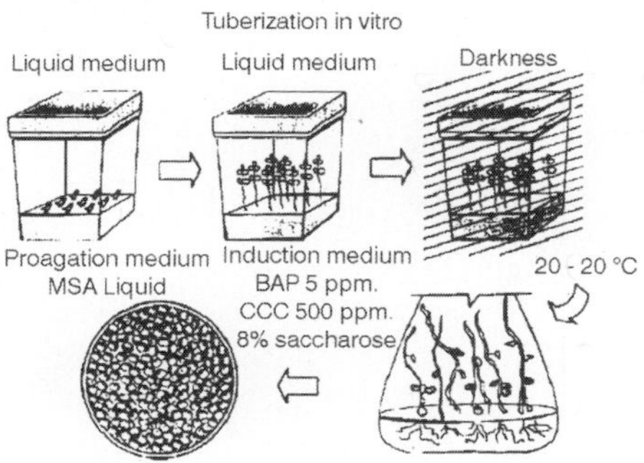

Fig. Poato in Vitro Tuberization Process Scheme

In the beginning, microtubers were used as an alternative for germplasm distribution and in vitro conservation. Curing the plantlet in vitro multiplication process, usually there is a multiplication of higher amounts to those required in the greenhouse and so, after the transference, some magentas are left over. These plantlets may be used for microtuber induction. In accordance with the indicated procedures, the induction medium is added to the magentas and then they are placed in the dark room. After three months the microtubers will be produced. These may be harvested and transferred to sterile containers (at 4°C), where they can be maintained up to 10 months, or used in the next campaign instead of in vitro plantlets.

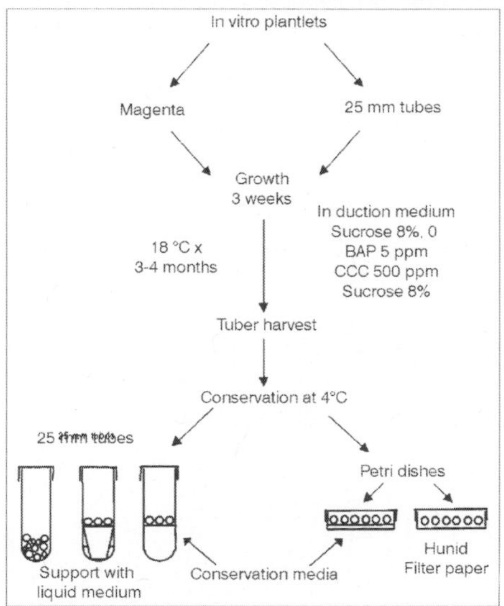

Fig. Potato in Vitro Tuberization Process Scheme

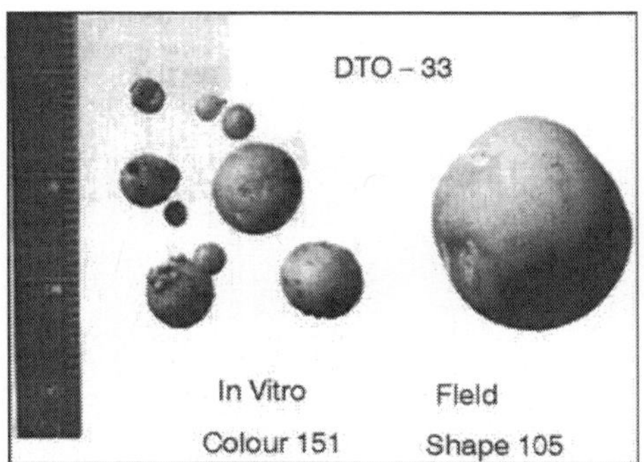

Fig. Comparison, in Size and Shape of in Vitro Microtubers with a Normal one from the Field

These microtubers can also be used as a reserve, in case in vitro material is contaminated or dies, because of temperature or handling effects. The high number and bigger size of the microtubers increases the success rates for the transfer to the greenhouse.

MEDIUM TERM GERMPLASM MAINTENANCE

The in vitro germplasm maintenance under normal growth conditions requires a series of transfers of the plantlets into a fresh medium. This leads to a consumption of time, increases the possibility of losses because of material contamination during successive sub-cultures, causes a loss of material by human error or failure of some equipment, and demands more labour A way to avoid these problems is through limitation, restriction or inhibition of growth. This approach consists of growth speed reduction by modifying the physical or chemical conditions of the culture, and it is effective for a short or medium term period.

In Vitro Conservation Methods

These consist in maintaining the cultures (buds, plantlets derived from nodes or directly from meristems) under physical (environmental factors] or chemical (culture medium composition) stressed conditions that make it possible to extend, as much as possible, the interval of transference into the fresh media, without affecting the viability of the cultures. The methods to reduce the in vitro cultivated plant growth include the reduction of temperature and light during storage, the incorporation of growth retardants in the medium, and the induction of osmotic stress in the medium, or a combination of all these.

Temperature

Temperature reduction has been the most used way to curb culture growth. Most of the in vitro cultures are maintained at temperatures between 12 and 200C: at lower temperatures, the growth rate decreases but the reduction depends on the species.

Nutrients Concentration

The reduction of the carbohydrate concentration and the nitrogenous components of the nutritive medium may affect the growth rates. In addition, the continuous absorption of these nutrients during plantet growth will bring a nutritional defficiency, which could produce a premature death of the plants.

Use of Growth Regulators

The abscisic acid (ABA), phosphon-D, maleic hydrazide, and succinic acid are some of the most frequently used growth regulators.

Osmotic Concentration

Growth limitation caused by osmotic concentration is due to the reduction of the water and nutrients absorbed from the medium. For example, at high concentrations, sucrose acts osmotically and it is highly metabolized. No metabolized osmotics, such as Manitol and Sorbitol, are possibly more efective than sucrose in culture growth limitation.

Evaluation of in Vitro Preserved Material

To evaluate material under these conditions, some important facts about in vitro maintenance such as genetic viability and stability should be considered. The viability evaluation of the in vitro cultures must be systematic. In slow growth conditions, when the sub-culture or transfer period extends during months or years, the frequency of the culture evaluation increases. The most important characteristics to be evaluated in the low-growth cultivars storage of apical buds are: contamination, leaf senescence, the number of green sprouts, the number of viable nodes in relation to the stem length, the presence or not of roots, and callus formation.

MAINTENANCE OF ACCESSIONS IN SEED PROGRAMME

In a seed programme, a group of accessions free of virus is maintained for pre-basic seed production: however, most of them are not propagated in the greenhouse so they are maintained in vitro to be used in the future. The continuous propagation of the in vitro plantlets damages the material, mainly if we consider that environmental conditions are not adequate. The alternative is to maintain the accessions in conservation media. Each accession must be maintained in test tubes with five replications to avoid possible losses. The maintenance conditions have been tested in CIP's potato collection that consists of more than 5,000 accessions, by means of which it is possible to assure the normal recovery of the material after using stress producers.

The plantlets used in each campaign must come from the maintenance phase. Then, they will be propagated in normal media where their growth will be reestablished, and the elected procedures for plantlet multiplication in the greenhouse will continue.Each season must be initiated with this material to start off with strong plants. In addition, the plantlets taken for multiplication must be replaced, trying to maintain 5 tubes per accession in the conservation media. The sub-cultures in the conservation medium are carried out approximately every one or two years. The conservation medium renovation must pass through a previous sub-culture in the propagation media to rejuvenate the explants.

VIRUS ERADICATION THROUGH MERISTEMS CULTURE AND THERMOTHERAPY

If a healthy plant is sown in the field, it is exposed to infections caused by pathogens as nematodes, fungi, bacteria, phytoplasms, virus and viroids, which have a negative effect on yield, and in some cases may kill the plants. However, not all the plant cells may become infected. A group of cells, which are in a continuous non-differentiated multiplication, are virus-free: the meristem. The in vitro meristems culture, together with growth at high temperatures produces potato plantlets free from viruses in more than 90 per cent of planted meristems. This routine method was established in the International Potato Centre to obtain virus-free plantlets for national and international distribution. The in vitro maintenance of virus-free plants provides the possibility to maintain all the time a bank of healthy and more vigourous plants, and with a more accelerated growth, than the infected ones. In a seed programme, it is essential to initiate this work with virus-free plantlets since this will affect the seed quality and the yield as well. The virus cleaning procedure is too long and expensive: that is why CIP, through its germplasm distribution programme, has provided a list of the principal virus-free potato varieties to all users.

PROCEDURE

- Approximately 18 to 20 plantlets (virus-infected) are propagated in magentas.
- After a growth period of 20 to 25 days, or when the plantlets are 4-5 cm high, they are placed in the thermotherapy chamber. The growth conditions are: 16 hours of light 34°C; 8 hours of darkness 32°C
- The magentas are maintained in the thermotherapy chamber for one month.
- Afterwards, the magentas are taken out of the chamber, and the outside is cleaned with 980/o alcohol. Then they are introduced to the culture room.
- Meristems are obtained as follows:

Cut the apical portion and remove the leaves that cover the meristem (approximately 3 to 4 leaves); the meristem is observed with a prominent leaf primordium. Remove the meristem with part of the leaf primordium; cut only the translucent portion. Use a new knife.Place the menistem in the culture medium. Be sure that the meristem is in the tube.

- Evaluate meristem growth and transfer them to fresh media if necessary.
- Each meristem that originates a plant is called <'line», which will be labeled according to the accession it belongs to. For example, Yellow Line 1, Yellow Line 2, Yellow Line 3, etc.
- Five tubes with several plants are propagated to evaluate the potato virus PSTVd, PVT, to determine the host range, the morphologic evaluation, and the in vitro maintenance.
- The results of each evaluation are obtained, and the infected material is replaced with clean material.

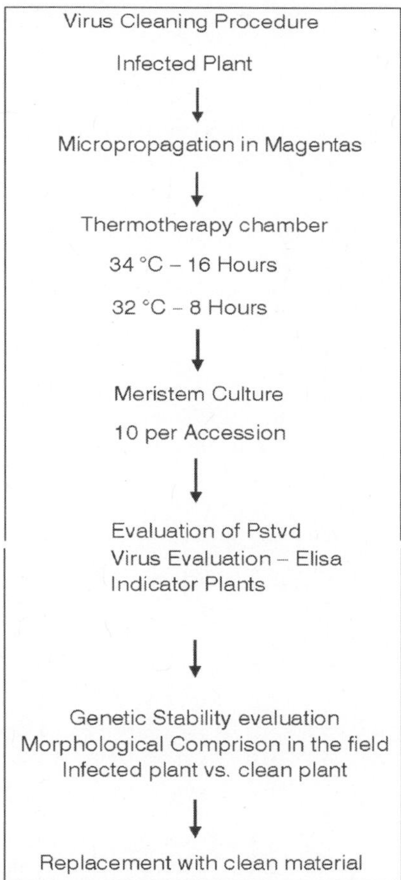

Temperature

The growth of the plantlets at temperatures higher than 20°C is accelerated (Do not use temperatures higher than 30°C).

CULTURED CELLS AND TISSUES

The reduction of the carbohydrate concentration and the nitrogenous components of thes pioneered in the USA. During the last thirty years, tissue culture-based plant propagation has emerged as one of the leading global agro-technologies. Between 1986 and 1993, the worldwide production of tissue cultured plants increased 50 per cent. In 1993, the production was 663 million plants. By 1997, production had risen to 800 million plants. During 1990–1994, the micropropagation industry declined in Europe, mainly due to production shifting to developing countries, but since then because of the demand for high quality and number, production in European countries has increased.

Since 1995, production has increased by 14 per cent in Asian countries, mainly due to the market entry of China, while the increase in South and Central America was from production in Cuba. More recently, some companies from Israel, the USA and UK have shifted their production requirements to Costa Rica and India.

Tissue cultured plants have as yet to reach many growers and farmers in the developing countries. The primary advantage of micropropagation is the rapid production of high quality, disease-free and uniform planting material. The plants can be multiplied under a controlled environment, anywhere, irrespective of the season and weather, on a year-round basis. Production of high quality and healthy planting material of ornamentals, and forest and fruit trees, propagated from vegetative parts, has created new opportunities in global trading for producers, farmers, and nursery owners, and for rural employment. Micropropagation technology is more expensive than the conventional methods of plant propagation, and requires several types of skills. It is a capital-intensive industry, and in some cases the unit cost per plant becomes unaffordable. The major reasons are cost of production and know-how. During the early years of the technology, there were difficulties in selling tissue culture products because the conventional planting material was much cheaper.

Now this problem has been addressed by inventing reliable and cost effective tissue culture methods without compromising on quality. This requires a constant monitoring of the input costs of chemicals, media, energy, labour and capital. In the industrialized countries, labour is the main factor that contributes to the high cost of production of tissue-cultured plants.

To reduce such costs, some steps can be partially mechanized, *e.g.* use of a peristaltic pump for medium dispensing, and of dishwashers for cleaning containers. In the less developed countries of Africa, Asia, and Latin America, where labour is relatively cheaper, consumables such as media, culture containers, and electricity make a comparatively greater contribution to production costs.

For example, the cost of medium preparation (chemicals, energy and labour) can account for 30–35 per cent of the micropropagated plant production.

However, automated production processes based on pre-sterilized membrane capsules, bioreactors, mechanized explant transfer, and container sealing are not commercially viable propositions in many developing countries. Therefore, low cost alternatives are needed to reduce production cost of tissue-cultured plants.

Many of the low cost technology options described in this publication can be incorporated in various steps of plant micropropagation. The occasional tissue culturegenerated variants (somaclones) and rare spontaneous bud mutants as well as those obtained from induced mutations can also be propagated by deployment of low cost techniques described. Micropropagation, in combination with radiation-induced mutations, speeds up the recovery, multiplication and release of improved varieties in vegetatively propagated plants. Hence, low cost technology will be of great value for large scale plant multiplication of mutants of many fruits, shrubs, flowers and forest trees that are conventionally vegetatively propagated.

PLANT TISSUE CULTURE

Plant tissue culture refers to growing and multiplication of cells, tissues and organs of plants on defined solid or liquid media under aseptic and controlled environment. The commercial technology is primarily based on micropropagation, in which rapid proliferation is achieved from tiny stem cuttings, axillary buds, and to a limited extent from somatic embryos, cell clumps in suspension cultures and bioreactors. The cultured cells and tissue can take several pathways. The pathways that lead to the production of true-to-type plants in large numbers are the preferred ones for commercial multiplication. The process of micropropagation is usually divided into several stages *i.e.*, prepropagation, initiation of explants, subculture of explants for proliferation, shooting and rooting, and hardening. These stages are universally applicable in large-scale multiplication of plants. The delivery of hardened small micropropagated plants to growers and market also requires extra care.

Plant tissue culture refers to growing and multiplication of cells, tissues and organs on defined solid or liquid media under aseptic and controlled environment. Plant tissue culture technology is being widely used for large-scale plant multiplication. The commercial technology is primarily based on micropropagation, in which rapid proliferation is achieved from tiny stem cuttings, axillary buds, and to a limited extent from somatic embryos, cell clumps in suspension cultures and bioreactors.

EXPLANT SOURCE

Plant tissue cultures are initiated from tiny pieces, called explants, taken from any part of a plant. Practically all parts of a plant have been used successfully as a source of explants. In practice, the "explant" is removed surgically, surface sterilized and placed on a nutrient medium to initiate the mother culture, that is multiplied repeatedly by subculture. The following plant parts are extensively used in commercial micropropagation. Shoot-tip and meristem-tip culture: Shoots develop from a small group of cells known as shoot apical meristem. The apical meristem maintains itself, gives rise to new tissues and organs, and communicates signals to the rest of the plant. Shoot-tips and meristem-tips are perhaps the most popular source of explants to initiate tissue cultures. The shoot apex explant measures between 100 to 500 μ m and includes the apical meristem with 1 to 3 leaf primordia. The apical meristem of a shoot is the portion lying distal to the youngest leaf primordium, and is ca.100 μ??m in diameter and 250μ?m in length with 800-1200 cells. In practice, shoot-tip explants between 100 to 1000 μ m are cultured to free plants from viruses. Even explants larger than 1000 μ m have been frequently used. The term "meristem-tip culture" has been suggested to distinguish the large explants from those used in conventional propagation.

Nodal or axillary bud culture: This consists of a piece of stem with axillary bud culture with or without a portion of shoot. When only the axillary bud is taken, it is designated as "axillary bud" culture. Floral meristem and bud culture: Such explants are not commonly used in commercial propagation, but floral meristems and buds can generate complete plants. Other sources of explants: In some plants, leaf discs, intercalary meristems from nodes, small pieces of stems, immature zygotic embryos and nucellus have also been used as explants to initiate cultures.

Cell suspension and callus cultures: Plant parts such as leaf discs, intercalary meristems,-stem-pieces, immature embryos, anthers, pollen, microspores and ovules have been cultured to initiate callus. A callus is a mass of unorganized cells, which in many cases, upon transfer to suitable medium, is capable of giving rise to shoot-buds and somatic embryos, which then form complete plants. Such calli on culture in liquid media on shakers are used for initiating cell suspensions. Liquid suspension cultures maintained on mechanical shakers achieve fast and excellent multiplication rates. However, in commercial micropropagation, calli are cultured mostly in bottles and flasks kept on semi-solid or liquid media. To a limited extent, bioreactors have become popular for somatic embryogenic cultures. It is considered that some day robotics could be adapted to bioreactorbased micropropagation.

PATHWAYS OF CULTURED CELLS AND TISSUES

The cultured cells and tissue can take several pathways to produce a complete plant. Among these, the pathways that lead to the production of true-to-type plants in large numbers are the popular and preferred ones for commercial multiplication. The following terms have been used to describe various pathways of cells and tissue in culture.

Regeneration and Organogenesis

In this pathway, groups of cells of the apical meristem in the shoot apex, axillary buds, root tips, and floral buds are stimulated to differentiate and grow into shoots and ultimately into complete plants. In many cases, the axillary buds formed in the culture undergo repetitive proliferation, and produce large number of tiny plants.

The plants are then separated from each other and rooted either in the next stages of micropropagation or in vivo (in trays, small pots or beds in glasshouse or plastic tunnel under relatively high humidity). The explants cultured on relatively high amounts of auxin (*e.g.* (2,4-D, 2,4-dichlorophenoxyacetic acid) form an unorganized mass of cells, called callus. The callus can be further sub-cultured and multiplied.

The callus shaken in a liquid medium produces cell suspension, which can be subcultured and multiplied into more liquid cultures. The cell suspensions form cell clumps, which eventually form calli and give rise to plants through organogenesis or somatic embryogenesis. In some cases, explants *e.g.* leaf-discs and epidermal tissue can also generate plants by direct organogenesis and somatic embryogenesis without intervening callus formation, *e.g.* in orchardgrass. Dactylis glomerata L.. In organogenesis the cultured plant cells and cell clumps (callus) and mature differentiated cells (microspores, ovules) and tissues (leaf discs, inter-nodal segments) are induced to differentiate into complete plants to form shoot buds and eventually shoots, and rooted to form complete plants.

Somatic Embryogenesis

In this pathway, cells or callus cultures on solid media or in suspension cultures form embryo-like structures called somatic embryos, which on germination produce complete plants. The primary somatic embryos are also capable of producing more embryos through secondary somatic embryogenesis. Although, somatic embryogenesis has been demonstrated in a very large number of plants and trees, the use of somatic embryos in large-scale commercial production has been restricted to only a few plants, such as carrot, date palm, and a few forest trees.

Somatic embryos are produced as adventitious structures directly on explants of zygotic embryos, from callus and suspension cultures. Somatic embryos and synthetic seeds (embryos encapsulated in artificial endosperm) hold potential for large-scale clonal propagation of superior genotypes of heterogeneous plants.They have also been used in commercial plant production and for the multiplication of parental genotypes in large-scale hybrid seed production. In many species, somatic embryos are morphologically similar to the zygotic embryos, although some biochemical, physiological and anatomical differences have been documented.

The synthetic auxin, 2,4-D is commonly used for embryo induction. In many angiosperms, *e.g.*, carrot and alfalfa, subculture of cells from 2,4-D containing medium to auxin-free medium is sufficient to induce somatic embryogenesis. The process can be enhanced with the application of osmotic stress, manipulation of medium nutrients, and reducing humidity. Selection of embryogenic cell lines has also been successfully used. For example, selection for unique morphotypes in grapevine cultures allows production of high quality embryos with predicable frequency.

A major problem in large-scale production of somatic embryos is culture synchronization. This is achieved through selecting cells or pre-embryonic cell clusters of certain size, and manipulation of light and temperature, temporary starvation or by adding cell cycle synchronizing chemicals to the medium. Cytokinins seem to play a key role in cell cycle synchronization and embryo induction, proliferation and differentiation. Abscisic acid is crucial in all the stages of somatic development, maturation and hardening.

Synthetic Seeds

The concept of production and utilization of synthetic seeds (somatic embryo as substitutes for true seeds) was first suggested by Murashige in 1977. Synthetic seeds can be produced either as coated or non-coated, desiccated somatic embryos or as embryos encapsulated in hydrated gel (usually calcium alginate).Successful utilization of synthetic seeds as propagules of choice requires an efficient and reproducible production system and a high percentage of post-planting conversion into vigourous plants. Artificial coats and gel capsules containing nutrients, pesticides and beneficial organisms have long been thought as substitutes for seed coat and endosperm. However, this technology is still in the developmental stage, and currently cannot compete with the other methods of commercial plant propagation.

Process of Micropropagation

The process of plant micro-propagation aims to produce clones (true copies of a plant in large numbers).

The process is usually divided into the following stages:

Stage 0 – Pre-propagation step or selection and pre-treatment of suitable plants.

Stage I – Initiation of explants-surface sterilization, establishment of mother explants.

Stage II – subculture for multiplication/proliferation of explants.

Stage III – shooting and rooting of the explants.

Stage IV – Weaning/hardening.

These stages are universally applicable in large-scale multiplication of plants. The individual plant species, varieties and clones require specific modification of the growth media, weaning and hardening conditions. A rule of the thumb is to propagate plants under conditions as natural or similar to those in which the plants will be ultimately grown ex-vitro. For example, if a chrysanthemum variety is to be grown under long day-length for flower production, it is better to multiply the material under long-day length at stages III and IV. There is a wide option to undertake production of plant material up to a limited number of stages. For example, many commercial tissue culture companies undertake production up to Stage III, and leave the remaining stages to others.

Pre-Propagation Stage

The pre-propagation stage (also called stage 0) requires proper maintenance of the mother plants in the greenhouse under disease- and insect-free conditions with minimal dust. Clean enclosed areas, glasshouses, plastic tunnels, and net-covered tunnels, provide high quality explant source plants with minimal infection. Collection of plant material for clonal propagation should be done after appropriate pretreatment of the mother plants with fungicides and pesticides to minimize contamination in the in vitro cultures. This improves growth and multiplication rates of in vitro cultures.

The control of contamination begins with the pretreatment of the donor plants. They may be prescreened for diseases, isolated and treated to reduce contamination. The explants are then brought to the production facility, surface sterilized and introduced into culture. They may at this stage be treated with antibiotics and fungicides as well as anti-microbial formulations, such as PPM. The explants are then culture indexed for contamination by standard microbiological techniques, which are occasionally supplemented with tests based on molecular biology or other techniques.

Stage I

This stage refers to the inoculation of the explants on sterile medium to initiate aseptic culture. Initiation of explants is the very first step in micropropagation. A good clean explant, once established in an aseptic condition, can be multiplied several times; hence, explant initiation in an aseptic condition should be regarded as a critical step in micropropagation. More than often, explants fail to establish and grow, not due to the lack of a suitable medium but because of contamination. The explants are transferred to in vitro environment, free from microbial contaminants. The process requires excision of tiny plant pieces and their surface sterilization with chemicals such as sodium hypochlorite, ethyl alcohol and repeated washing with sterile distilled water before and after treatment with chemicals. After a short period of culture, usually 3 to 5 days, the contaminated explants are discarded.The surviving explants showing growth are maintained and used for further subculture. In herbaceous plants *e.g.* potato, chrysanthemum, carnation, streptocarpus, strawberry, and African violet; the explant sources are meristems, apical- and axillary buds, young seedlings, developing young leaves and petioles, and unopened floral buds. The following low cost options can be adapted to initiate explants:

Sterile Instrument Technique

This method assumes that most of the deep-seated meristems and those covered by leaves or other integuments (*e.g.* floral bracts) are sterile. In this procedure, the explant is washed with sterile water, rinsed in ethanol, and instruments are sterilized every time they touch the surface of the explant, and the explant is moved to a new location on the dissection stage.

Surface Sterilization Technique

This is by far the most commonly used method. The explants are washed in sterile water, rinsed in ethanol, and surface sterilization is achieved by using chemicals with chlorine base. Calcium or sodium hypochlorite based solutions, 1–3 per cent (v/v) are usually used for soft herbaceous materials. A cheap and ready-made sterilant is 5-7 per cent solution of 'Domestos'- a toilet disinfectant which contains 10.5 per cent v/v sodium hypochlorite, 0.3 per cent sodium carbonate, 10.0 per cent sodium chloride and 0.5 per cent (w/v) sodium hydroxide and a patented thickener). The explants are washed in sterile distilled water before and after sterilization. Other surface sterilants used include mercuric chloride (avoid its use as far as possible, since it is highly toxic), hydrogen peroxide, and potassium permanganate.

For Soft Tissues

- Wash explants from perennial plants for 1-2 hr in tap water. Eliminate this step for material from glasshouse grown plants.
- Wash in sterile distilled water three to four times for 5 to 10 minutes each.
- Dip in 95 per cent ethanol for 3 to 5 seconds.
- Wash once again with sterile distilled water for 5 minutes.
- Surface-sterilize in 5 per cent 'Domestos' (v/v) for 20-25 minutes.
- Wash with sterile distilled water three times for 10 minutes each.
- Drain water droplets by placing on pre-sterilized blotting paper.
- Transfer explants singly to the medium.

N.B.: Sterilize forceps each time to transfer explants to avoid cross-contamination.

For woody stems (e.g. roses, hardy shrubs, and trees):

- Collect stems, shoots, buds and store at 5 oC till needed.
- Rinse in ethanol for 3 to 5 seconds.
- Rinse in 1- per cent sodium hypochlorite (20 per cent bleach) for 10 minutes.

- Place lower parts of stems in flasks in 2 per cent sucrose and 200 PPM 8-hydroxyquinoline citrate at 23+2 oC. For items collected in September/October, add 50-PPM GA3.

After that 10 PPM GA will help break the dormancy:

- Re-cut the bottom of stem and replace the solution after 2 days.
- Excise the softwood from the developed shoots and use material for explants or for rooting.
- Surface-sterilize as in the above protocol.

Do not forget to sterilize forceps and scalpel every time for the transfer of explants to fresh solutions. Use sterile containers in the protocol of surface sterilization. If explants become brown or pale at the end of the protocol, reduce the strength of 'Domestos' to 2.5 per cent. Alternatively, dip explants in 10 per cent 'Domestos' for 2 minutes and then proceed to surface sterilize with 3-5 per cent 'Domestos' for 20 minutes. If basal contamination is observed after 2-3 days of culture, explants can sometimes be rescued by removing the basal end by making a single cut with a sharp scalpel and re-culturing on fresh medium.

Stage II

Stage II is the propagation phase in which the explants are cultured on the appropriate media for multiplication of shoots. The primary goal is to achieve propagation without losing the genetic stability. Repeated culture of axillary and adventitious shoots, cutting with nodes, somatic embryos and other organs from Stage I leads to multiplication of propagules in large numbers. The propagules produced at this stage can be further used for multiplication by their repeated culture. Sometimes it is necessary to subculture the in vitro derived shoots onto different media for elongation.

Stage III

The in vitro shoots obtained at Stage II are rooted to produce complete plants. If the proliferated material consists of bud-like structures (*e.g*. orchids) or clumps of shoots (banana, pineapple), they should be separated after rooting and not before. Many plants (*e.g*. banana, pineapple, roses, potato, chrysanthemum, strawberry, mint, several grasses and many more) can be rooted on half-strength-MS medium without any growth-regulators. Good sturdy well-rooted plants are essential for high survival during weaning and later transfer to soil. This stage is labour intensive and expensive. The process of in vitro rooting has been estimated to account for approximately 35–75 per cent of the total cost of production. Efforts should be made to combine rooting and acclimatization stages.

Stage IV

At this stage, the in vitro micropropagated plants are weaned and hardened. This is the final stage of the tissue culture operation after which the micropropagated plantlets are ready for transfer to the greenhouse. Steps are taken to grow individual plantlets capable of carrying out photosynthesis. The hardening of the tissue-cultured plantlets is done gradually from high to low humidity and from low light intensity to high intensity conditions. If grown on solid medium, most of the agar can be removed gently by rinsing with water.

Plants can be left in shade for 3 to 6 days where diffused natural light conditions them to the new environment. The plants are then transferred to an appropriate substrate (sand, peat, compost, etc.), and gradually hardened. Low-cost options include the use of plastic domes or tunnels, which reduces the natural light intensity and maintains high relative humidity during the hardening process. If the plants are still joined together after rooting, these should be planted as bunches in the soil and separated after 6 to 8 weeks of growth.

DELIVERY TO THE GROWERS

The delivery of the rooted and hardened small micropropagated plants to growers and market requires extra care. In some cases, plant losses can occur during shipment and handling by growers. This is particularly true when the plants are not fully hardened and rooted or not grown for sufficient duration after transfer to soil. Growers should be given clear instructions how to handle the material provided. Apart from the economic loss, poor survival of planted material erodes the confidence of growers in the technology. The transfer of individual plants to soil in black plastic or polythene bags is widely used as a low-cost option to provide fully-grown banana plants directly to farmers in many developing countries.

CONCLUSION

Fungi and bacteria grow on the plant surface and contaminate the culture media when they are not adequately eliminated. The process of explant introduction into in vitro conditions depends mainly on the disinfecting phase. Problems in handling produce contamination of the in vitro plant and so, a similar disinfection process to that of the in vitro introduction technique is recommended.

However, it is possible to use antibiotics included in the medium, but just temporarily, while the plantlets are growing. Among the bacterial contaminants are Bacillus sp., Erwin/a sp., Pseudomonas sp., etc. In the case of systemic infections, the use of disinfectants is not effective, because the pathogens are located in the vascular system. In this case, it is recommended to use specific antibiotics and meristems or bud cuttings.

On the other hand, the process of in vitro introduction involves the use of explants in different physiological stages, as dormant buds, which require growth regulators for example, Gibberellic acid) to stimulate and accelerate bud growth.

During the process of in vitro introduction there is a close connection between the greenhouse and the laboratory; that is why the necessary precautions should be taken to avoid the entrance of contaminants into the laboratory.

- In the greenhouse
  - Take cuttings from the mother plant, or buds from greenhouse tuber sprouts.
- In the laboratory
  - Immerse the cuttings for 10 minutes in a beaker conveniently labeled with the accession name, and containing a solution of 0.5 per cent acaricide with three drops/I of Tween 20.
  - Throw away the acaricide solution and rinse the nodes with tap water
  - Prepare the laminar flow chamber.
  - Immerse the cuttings in 700/o alcohol for 30 seconds and take the beakers immediately to the laminar flow chamber
- In the laminar flow chamber
  - Eliminate the alcohol and replace it with a 2.50/o solution of calcium hypochlorite. Keep the cuttings immersed during 15 minutes.
  - Throw away the hypochlorite, and rinse the cuttings three times with sterile water
  - Keep the nodes immersed in sterile water until bud extraction.
  - Dissect the buds on a sterile plate and place them in the culture medium.

7

Quantitative Genetic Theory

THE ESTIMATION OF QUANTITATIVE GENETIC PARAMETERS

The main activity of a plant breeder does not consist of making quantitative genetic studies of a number of traits, but the development of new varieties. This means that breeders are unwilling to dedicate great efforts to the estimation of quantitative genetic parameters. Thus only estimation procedures demanding hardly any additional effort, fitting in a regular breeding programme, are presented in this section.

First attention is given to some problems involved in obtaining appropriate estimates of var(e), the environmental variance. Because of these problems, in the present section procedures for estimating var(G) or $h2$ not requiring estimation of var(e) are emphasized.

Breeders may measure the phenotypic variation for a trait of some genetically heterogeneous population. They may do so by estimating var(p). However, their main interest lies in exploiting the genetic variation. As

$$\text{var}(\underline{G}) = \text{var}(\underline{P}) - \text{var}(\underline{e})$$

an appropriate way to estimate var(G) consists of subtracting v^ar(e) from v^ar(p).

The estimate for var(e) should be derived from similar but genetically homogeneous plant material, grown in the same macro-environmental conditions as the population of interest. A complication arises if the genotypes differ in their capacity to buffer variation in the growing conditions. Then the candidates representing one genotype are more (or less) affected by the prevailing variation in the quality of the micro-environmental growing conditions than the candidates plants representing another genotype. This was already dealt with in Example and its preceding text.

To account for this, the environmental variance assigned to the F2 population of a self-fertilizing crop is sometimes estimated to be:

$$\frac{1}{4}\hat{var}(\underline{P}_{P1}) + \frac{1}{2}\hat{var}(\underline{P}_{F1}) + \frac{1}{4}\hat{var}(\underline{P}_{P2})$$

Plants of the F2 generation are more heterozygous than those of P1 or P2, but less than those of the F1. Heterogeneity among plants of the F1 may be partly due to the manipulations applied to produce the F1 seed, *i.e.* emasculation and pollination of the parent (instead of spontaneous selfing). Manipulation certainly contributes to heterogeneity in the case of cloning. Thus the usual way of cloning (*e.g.* of grass or rye plants) gives clones such that the withinclone phenotypic variance overestimates the environmental variance appropriate to the segregating plant material not subjected to the manipulation required for the cloning. Example illustrates the present concern of using a non-representative estimate of var(e).

POPULATION GENETIC AND QUANTITATIVE GENETIC EFFECTS OF SELECTION BASED ON PROGENY TESTING

Section introduced the concept of breeding value as a rather abstract quantity applying in the case of random mating. In Section it was emphasized that the concept is of great importance when selectingamongcandidatesonthe basisprogenytesting.Thepresent sectionaims to clarify population genetic and quantitative genetic effects of such selection. The progenies to be evaluated are obtained by crossing of candidates with a so-called tester population. In Section it was shown that, in the case of selfing, haplotype frequencies hardly change in course of the generations.

Thus it does not matter so much whether one evaluates the breeding value of individual plants or the breeding value of lines derived from these plants. The obtained progenies are HS-families.

The tester population may be

- The population to which the candidates belong (intrapopulation testing).
- Another population (interpopulation testing).

Intrapopulation Testing

In the case of intrapopulation testing the allele frequencies of the tester population are equal to the allele frequencies of the population of candidates: p and q. Open pollination, as in the case of a polycross, is of course the simplest way of obtaining the progenies.

Interpopulation Testing

When applying interpopulation testing, the tester population is another population than the population of candidates. Its allele frequencies are designated $p\_$ and $q\_$. The aggregate of all families resulting from the test-crosses is then equal to the population resulting from bulk crossing. Interpopulation testing occurs at top-crossing and at reciprocal recurrent selection. In top-crossing a set of (pure) lines, which have been emasculated, are pollinated by haplotypically diverse pollen, possibly produced by an SC-hybrid or by a genetically heterogeneous population. At so-called early testing, young lines are involved in the top-cross.

With regard to the candidates being tested, we now consider

- The effect of the allele frequencies in the tester population on the ranking of the candidates with regard to their breeding value
- The effect of selection of candidates with a high breeding value on the allele frequencies and, as a consequence, the expected genotypic value

THE EFFECT OF THE ALLELE FREQUENCIES IN THE TESTER POPULATION ON THE RANKING OF THE CANDIDATE GENOTYPES WITH REGARD TO THEIR BREEDING VALUE

When selecting (parental) plants with regard to their breeding values, plants with the most attractive (possibly: the highest) breeding values are selected. However, the ranking of the breeding values of plants with genotype *bb, Bb* or *BB* is not straightforward. It depends on the frequency of allele *B* in the tester population. This complicating factor is now considered. The selection among the candidates is based on the quality of their offspring, *i.e.* on their breeding value. Table shows that, for a given allele frequency (*p*), the ranking of the candidates with regard to their breeding value depends on whether *á*_ is positive, zero or negative. The ranking depends thus on whether

$$a' = a - (P' - q')d = a - (2p' - 1)d = (a + d) - 2p'd$$

is positive, zero or negative. This depends for a given locus, *i.e.* for given values for *a* and *d*, on *p*_, the gene frequency in the tester population. The values for *p*_ making *á*_ either positive, or zero or negative will now be derived. Because of the tendency that *d e*– 0 for most of the loci, these values will only be derived for loci with *d e*– 0. When considering Equation it is easily derived that

The reader is reminded that *pm* is the allele frequency giving rise to the maximum of E*G* in the case of the Hardy–Weinberg genotypic composition. At $d = a$ it amounts to 1, whereas $d > a$ implies $0 < pm < 1$.

Ranking of the candidate genotypes for increasing breeding value, *i.e.* increasing value for

Yields thus

$$if\ \alpha' > 0$$

$$b_{vbb} < b_{vBb} < b_{vBB},\ or: b_{v0} < b_{v1} < b_{v2}$$

$$if\ \alpha' = 0$$

$$b_{v0} = b_{v1} = b_{v2}$$

Ranking is impossible for loci with $d \geq a$, *if* $p' = P_m$

$$if\ \alpha' < 0$$

$$b_{v2} < b_{v1} < b_{v0}$$

In Section it was shown how one might estimate var($b\acute{\imath}$) = $\sigma^2 a$. In the case of a high value for var($b\acute{\imath}$) prospects for successful selection are good. One may help achieve that by using an appropriate tester population as well as uniform environmental conditions in the progeny test. The choice of the tester is especially relevant for loci with overdominance or pseudo-overdominance.

One should avoid using, with respect to such loci, a tester with $p\_\ H\!-pm$, as such a tester would yield equivalent progenies. Figure shows that $\acute{a}\_$, and consequently var($b\acute{\imath}$), is smaller as $p\_$ approaches either 1 or pm. The former concerns loci with (in)complete dominance, the latter loci with overdominance. In both these cases the tester population will have a high expected genotypic value.

In practice it has often been observed that σa^2 does not decrease when applying selection.

The Effect of selection of candidates with a High Breeding Value on the Expected Genotypic Value

In the context of progeny testing, the goal of the selection of candidates with a high breeding value is improvement of the genotypic value expected for the population subjected to the selection. It will be shown that this goal can not always be attained.

When combining the preceding text and the implications of Figure, it can be deduced that selection of candidate plants with a high breeding value implies

- If $a\_ > 0$ An increase of p. This is associated with an increase of EG if $0 \leq d \leq a$, or if $d > a$ as long as $p < p_m$. It is associated with a decrease of EG if $d > a$ and $p > p_m$.
- If $a\_ = 0$ No change in p, *i.e.* no change in EG.
- If $a\_ < 0$

A decrease of p. This is associated with an increase of EG as long as $p > pm$. It is associated with a decrease of EG if $p < p_m$.

It is assumed that absence of overdominance is the rule. The usual situation of presence of partial dominance or additivity, *i.e.* $0 \leq d \leq a$, implies then preferential selection of plants with genotype *BB*, *i.e.* an increase of p until p = 1. This is associated with an increase of EG. For the relatively rare loci with overdominance ($d > a$) three situations concerning the tester population, namely $p\_ = p_m$, $p\_ < p_m$ and $p\_ > p_m$, have to be distinguished:

- $p\_ = p_m$

A tester population with $p\_ = p_m$ prohibits meaningful progeny testing for the involved loci: the progeny test does not allow successful selection among the candidates with regard to their breeding values.

- $p\_ < p_m$

In this case the tester produces pollen with haplotype b in such a frequency that candidates with genotype *BB* tend to yield superior offspring, if indeed d

$> a$. Such candidates will be selected on the basis of the progeny test. The frequency of gene B will consequently increase.

- $p_- > p_m$

When using a tester population with $p_- > p_m$, candidates with genotype *bb* tend to produce superior offspring. Selection on the basis of the progeny test implies then a decrease of the frequency of allele B.

The above three situations for loci with overdominance require a more detailed treatment, both for

- Intrapopulation progeny testing and for
- Interpopulation progeny testing.

Intrapopulation Progeny Testing

Figure illustrates how the allele frequency p will change, starting from the initial value p_0, in the case of continued selection of candidates with a high breeding values.

This is done for a locus with $p0 > pm$ as well as for a locus with $p_0 < p_m$. The actual value of p_m depends, of course, on the values for a and d of the considered locus.

In both cases p approaches pm asymptotically. The closer pm is approached, the smaller the differences in breeding and the smaller the heritability, *i.e.* the less efficient the selection. The changes in p become then smaller.

At $p = p_m$ all genotypes have the same breeding value. In that situation the expected genotypic value (EG) is maximal. Further improvement is then impossible.

Figure depicts the same initial situation. Now, however, it is assumed that the selection results immediately in gene fixation, *i.e.* in $p_1 = 0$ (if $p_0 > pm$) or in $p_1 = 1$ (if $p_0 < p_m$). This may occur when selecting only a few candidate genotypes on the basis of testing progenies obtained from a polycross.

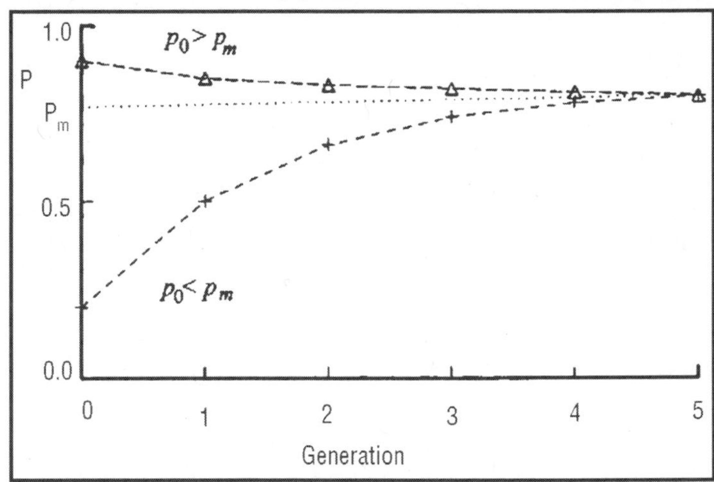

Fig. The presumed frequency of allele B in successive generations with selection, based on intrapopulation testing, of candidates with a high breeding value; for a locus with as well as a locus with in the case of continuous change of P.

If the aim is to develop a synthetic variety the result may be disappointing: the maximum value for E*G* will never be attained.

Still another possibility is that selection starting with $p_0 < p_m$ gives successively rise to $p_1 > p_m$, $p_2 < p_m$, $p_3 > p_m$, *etc.* (or that selection starting with $p_0 > p_m$ gives successively rise to $p_1 < p_m$, $p_2 > p_m$, $p_3 < p_m$, *etc.*). Then p oscillates around *pm*. Notwithstanding the presence of genetic variation the selection results in at most a small progress of E*G*, associated with dampening of the oscillation.

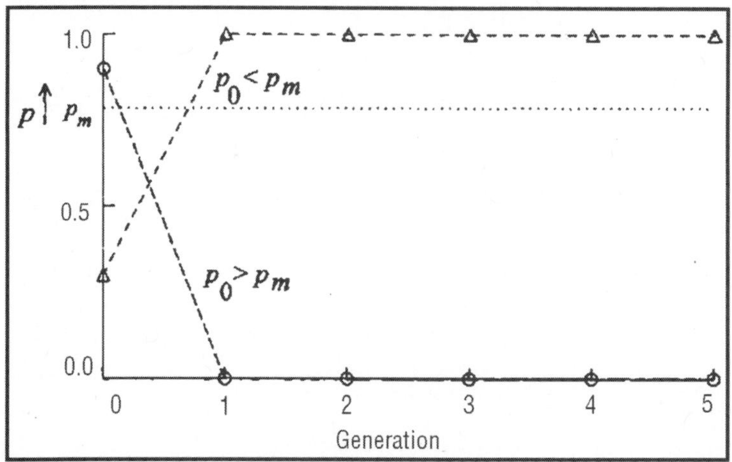

Fig. The presumed frequency of allele B in successive generations when selecting, based on intrapopulation testing, candidates with a high breeding value for a locus wit®h as well as a locus with in the case of fixation after selection in generation.

Interpopulation Progeny Testing

Interpopulation progeny testing occurs when applying recurrent selection (for general combining ability or specific combining ability) or reciprocal recurrent selection. In this paragraph attention is focussed on reciprocal recurrent selection (RRS). In RRS two populations, say A and B, are involved.

Plants in population A are selected because of their breeding values when using population B as tester. Likewise, and simultaneously, plants in population B are selected because of their breeding values when using population A as tester. (In an annual crop such as maize the S1 lines obtained from the plants appearing to have a superior breeding value are used to continue the programme.)

It is likely that the allele frequencies of populations A and B differ more as these populations are less related. If indeed the allele frequencies are very different, it is probable that $p_A > p_m > p_B$, or – at a different labelling of the populations – that $p_A < p_m < p_B$, where p_A designates the allele frequency in population A and p_B the allele frequency in population B. The first situation implies testing of candidates representing population A with a population with p_B such that

$$\alpha' = (a+d) - 2_{pBd} > 0$$

Selection in population A will then tend to yield an increase of pA. It also implies testing of candidates representing population B with a tester with pA such that $á\_ < 0$. Selection in population B tends then to yield an decrease of p_B. These tendencies are illustrated in Figure. Continued selection will then, eventually, yield the desired goal, *viz.* two populations mutually adapted such that a bulk cross between them yields, with regard to loci affecting the considered trait and with $d > a$, exclusively heterotic, heterozygous plants.

Figure depicts the development of the allele frequencies if the initial value of p_A is equal to pm. This implies for the candidates genotypes in population B that $á\_ = 0$. Effective selection of candidates with a high breeding value is then impossible in population B. The results eventually obtained is, however, the same as in Figure. This may even occur if $p_A < pm$ and $p_B << p_A$. Then, due to the first cycle of reciprocal recurrent selection, p may be increased in both populations such, that $p_A > pm$ and $p_B < p_m$. To help ensure that populations A and B have very different allele frequencies with regard to a large number of loci with $d > a$, these populations may be chosen on the basis of an evaluation of the performance of plant material produced by bulk crossing of a number of populations. Eligible populations are: open pollinating varieties, synthetic varieties, DC- TC- and SC-hybrid varieties. If for a certain locus p_A and p_B are very similar, interpopulation.

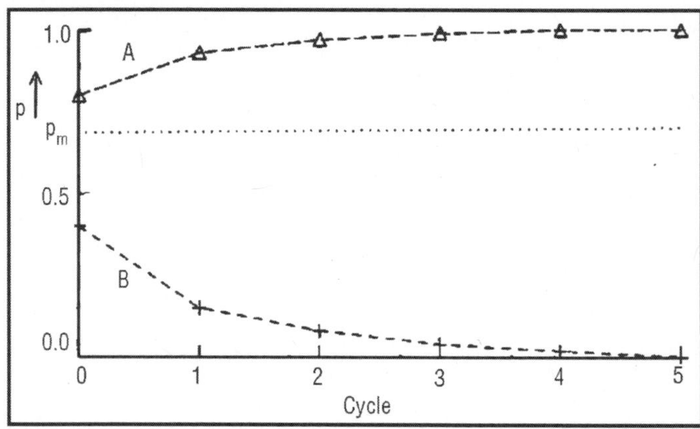

Fig. The Presumed Frequency of Aallele *B* in Successive Cycles of Reciprocal Recurrent Selection in Populations *A* and *B*, for a Locus with an Initial Allele Frequency (p_0) such that $p_0 > p_m$ in Population *A* and $p_0 < p_m$ in Population *B*.

Progeny testing resembles intrapopulation progeny testing. The selection will then, in both populations, induce p to approach pm. (This is illustrated in Figure for pA $H-$ pB, where both are less than pm). The result of continued selection will then be two populations with the Hardy-Weinberg genotypic composition, thus two populations with E*G* being equal to its maximum, *i.e.*

$$m + \frac{a^2 + d^2}{2d}$$

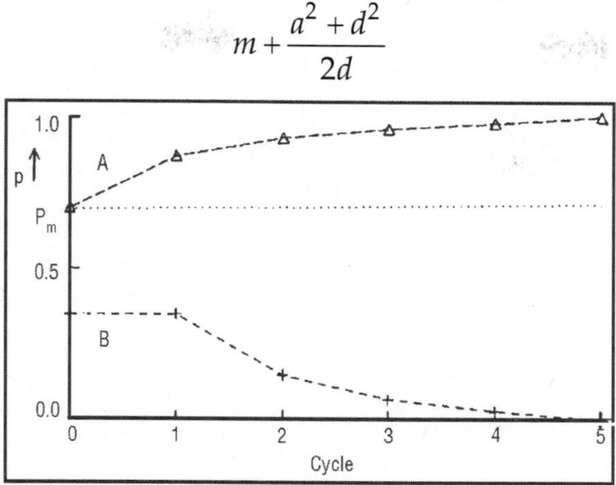

Fig. The presumed frequency of allele B in successive cycles of reciprocal recurrent selection in populations A and B, for a locus with an initial allele frequency (p_0) such that $p_0 = p_m$ in population A and $p_0 < p_m$ in population B.

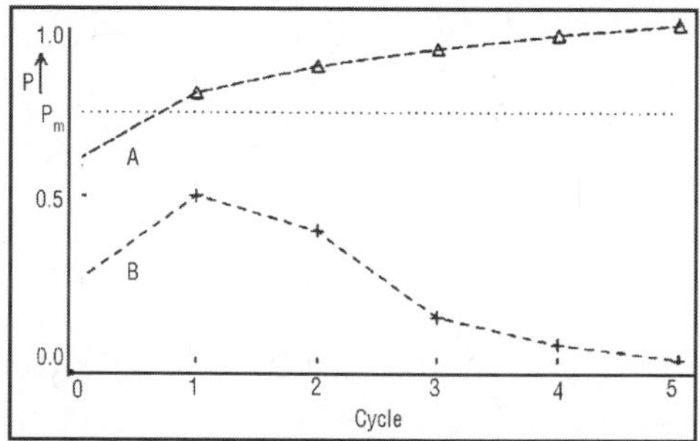

Fig. The presumed Ffrequency of Allele B in Successive Cycles of Reciprocal Recurrent Selection in Populations A and B, for a locus with strongly different initial allele frequences (but both smaller than P_m).

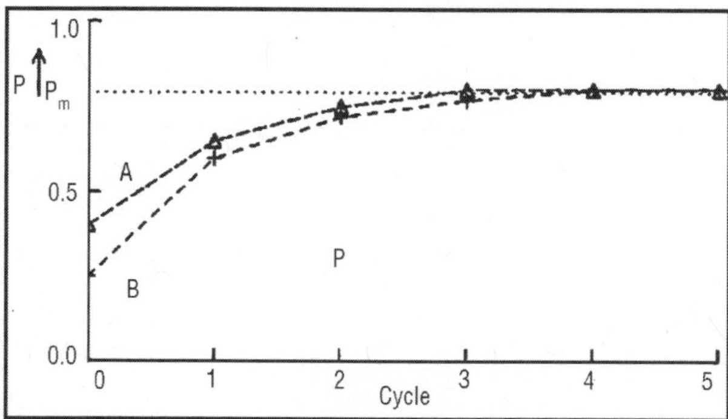

Fig. The presumed frequency of allele B in successive cycles of reciprocal recurrent selection in populations A and B, for a locus for which the initial allele frequencies are very similar.

For loci with $d > a$ this maximum is less than $m + d$. The ultimate goal of reciprocal recurrent selection is plant material obtained by a bulk cross of the improved populations. The expected genotypic value of that plant material is, due to the presence of genetic variation, less than the highest possible genotypic value $m+d$, *i.e.* the genotypic value of the heterotic heterozygous genotype.

PLANT MATERIAL WITH IDENTICAL REPRODUCTION

Clones, pure lines and single-cross hybrids can be reproduced with the same genotype. For such plant material, estimation of the heritability in the wide sense may proceed as elaborated in this section.A random sample consisting of *I* genotypes is taken from a population of entries with identical reproduction; $I > 1$. Each sampled genotype is evaluated by growing it in *J* plots, each containing *K* plants; $J > 1$,*K e*– 1. These plots may be assigned to

- A completely randomized experiment.
- Randomized (complete) blocks.

Table presents the analysis of variance for either design. The test of the null hypothesis H0: –σ*g* 2 = 0– requires calculation of the *F* value, MS*g*/MSr. This value is compared with critical values tabulated for different levels of significance.

Unbiased estimates of σ2 and σ*g*2 are

$$\hat{\sigma}^2 = MS_r$$

$$\hat{\sigma}_g^2 = \frac{MS_g - MS_r}{J}$$

Table: The structure of the analysis of variance of data obtained from I genotypes evaluated at J plots

(a) Completely randomized experiment

source of var *iation*	*df*	*SS*	*MS*	*E(MS)*
Genotypes	$1-1$	SS_g	MS_g	$\sigma^2 + J\sigma_g^2$
Re*sidual*	$1(J-1)$	SS_r	MS_r	σ^2

(b) RandomiZed complete block design

source of var *iation*	*df*	*MS*	*E(MS)*
Blocks	$J-1$	MS_b	$\sigma^2 + I\sigma_b^2$
Genotypes	$1-1$	MS_g	$\sigma^2 + J\sigma_g^2$
Re*sidual*	$(J-1)(I-1)$	MS_r	σ^2

For each entry the mean phenotypic value calculated across the *J* plots constitutes the basis for the decision to select it or not. Thus the appropriate environmental variance when testing each genotype at each of *J* plots is

$$\sigma_e^2 = \frac{\sigma^2}{J}$$

The wide sense heritability is thus

$$h_w^2 = \frac{\sigma_g^2}{\sigma_g^2 + \sigma_e^2} = \frac{\sigma_g^2}{\sigma_g^2 + \frac{\sigma^2}{J}}$$

It should be noted that substitution of the unbiased estimates for σe 2 and for σ_g^2 in Equation does not yield an unbiased estimate for *hw* 2.

The intention of replicated testing of entries in several plots is a reduction of the environmental variance. This induces the heritability to be higher at higher values for *J*. The ratio

$$\frac{hJ^2}{h_1^2}$$

i.e. the heritability when testing each entry in several plots to the heritability when testing each entry at a single plot, is now considered.

In doing so, in the remainder of this section symbols with the subscript 1 refer to non-replicated testing (*J* = 1), and symbols with the subscript *J* to replicated testing (*Je*– 2). The heritability appropriate when testing each entry at each of *J* plots is thus designated by

$$hJ^2 = \frac{\sigma_g^2}{\sigma_J^2}$$

where σ*J*² represents the phenotypic variance of the means of the entries across *J* plots, *i.e.*

$$\sigma_J^2 = \sigma_g^2 + \left(\frac{\sigma^2}{J}\right)$$

$$h_1^2 = \frac{\sigma_g^2}{\sigma_g^2 + \sigma^2} = \frac{\sigma_g^2}{\sigma_1^2}$$

$$\sigma_g^2 = h_1^2 \sigma_1^2$$

$$\sigma^2 = \sigma_1^2 - \sigma_g^2 = \sigma_1^2 - h_1^2 \sigma_1^2$$

$$\sigma_J^2 = h_1^2 \sigma_1^2 + \frac{\sigma_1^2 - h_1^2 \sigma_1^2}{J}$$

Table: The ratio of the heritability when testing each entry at J plots to the heritability when testing each entry at a single plot($h_1^{2)}$ for several values for h_1^2 and J

	h_1^2				
J	0.1	0.2	0.3	0.4	0.5
2	1.82	1.67	1.54	1.43	1.33
3	2.50	2.14	1.88	1.67	1.50
4	3.08	2.50	2.11	1.82	1.60

$$\frac{\sigma_J^2}{\sigma_1^2} = h_1^2 + \left(\frac{1-h_1^2}{J}\right) = \frac{1+h_1^2(J-1)}{J}$$

From Equations and it follows that

$$\frac{h_J^2}{h_1^2} = \frac{\sigma_1^2}{\sigma_J^2} = \frac{J}{1+h_1^2(J-1)}$$

Table presents the ratio *hJ* 2/ *h*1 2 for several values for *h*1 2 and *J*.

Especially for a (very) low value for *h*1 2 application of additional replications may be rewarding because of the large (relative) increase of the heritability. The largest relative improvement occurs when applying $J = 2$ instead of $J = 1$. Thus potato breeders should consider a system where each first-yearclone is represented by 2 seed potatoes instead of only 1, which is customary).

As a general conclusion it is stated that replicated testing promotes the efficiency of selection. If the replicated testing involves different macroenvironments it gives an indication of the stability as well.

In Section attention is given to the optimum number of replications, say *Jopt*. It is the number of replications giving rise to the maximum response to selection at a fixed number of plots. The ratio *hJ* 2/*h*1 2 is shown to play a crucial role in the derivation of *Jopt*. In connection with the foregoing, we consider the ratio

$$\frac{\sigma_b^2}{\sigma_b^2 + \sigma_w^2}$$

σ_b^2 *represents the between – entry component of* variance *and*

σ_w^2 *the within – entry component of* variance.

The ratio may be considered if from each entry $J > 1$ observations are available. This occurs in perennial crops, such as apple and oil palm, when observing in successive years the yield per year of individual plants. The quantitative genetic interpretations of these components of variance are σw 2: environmental variance in course of time and σb 2: genetic variance + variance due to variation in permanent environmental conditions (because of the permanent position in the field).

In statistics the ratio is called intraclass correlation coefficient or repeatability. The numerator of the ratio tends to be larger than σg 2, which causes the ratio to be larger than hw^2. In certain situations estimation of $h2$ is not as easy as estimation of the repeatability. Then one may simply estimate the repeatability as this quantity indicates the upper limit of hw 2.

Observations repeated in the course of time do not only allow estimation of the repeatability or the heritability, they also indicate the stability, for instance the presence or absence of certain genotype × year interaction effects.

CHOICE OF PARENTS AND PREDICTION OF THE RANKING OF CROSSES

Prior to actual selection among evaluated candidates, the breeder selects among conceivable crosses. Parents will only be crossed if the progeny to be obtained are expected to be promising enough to be rewarding for the efforts of the crossing work. It is, of course, very attractive to be able to determine beforehand which crosses have the highest chance of producing a commercially desirable cultivar. This allows valuable time and efforts to be concentrated on crosses with a higher probability of producing desirable genotypes. A cross prediction method is, of course, only useful to a plant breeder if it is effective in handling large numbers of crosses.

Crops differ considerably with regard to the amount of work involved in a pollination. A single pollination of a cucumber flower, for instance, may yield hundreds of seeds. In contrast, the efforts required for the pollination of a single wheat ear, for instance, are considerable.

A single pollination requires emasculation, in time, of the flowers alongside the ear to be pollinated, bagging of the ear, collection of the pollen and its transfer to the stigma of the flowers to be pollinated, and bagging again. Additionally the breeder should administrate the parents involved in the pollination. All this work will, hopefully, result in only one seed per pollinated flower. It should be clear that it may be wise to consider seriously the crosses to be made.

Often crosses are made on the basis of implicit expert knowledge, but the choice may be supported by explicit information. Schut distinguished five sources of such information:

- Information about *the phenotypes* of the potential parents.
- Information about *the genotypes* of the potential parents with regard to traits with known genetic control.
- Information about *differences between the potential parents* with regard to:
- Their geographic origin.
- Their pedigrees.
- Their values for a set of traits.

The size of the difference is thought to indicate the number of heterozygous loci in the F_1. This number is, in its turn, thought to determine the heterosis in the F_1 and/or the genetic variance in the segregating generations. Crossing of distantly related lines with desired genotypic values for the relevant traits, which are due to different genotypes, is expected to increase the probability of transgression in the segregating populations.

Transgression occurs if the segregating population contains with regard to some trait one or more lines with a phenotypic value outside the range given by the parental phenotypic values.

Pedigree data offer an opportunity to calculate the degree of relatedness of related parents. Such data are, however, often incomplete or unreliable. The pedigree information can be quantified by a measure of relatedness of two potential parents, for instance by the coefficient of coancestry.

The traits information may concern:

- Agronomic traits.
- Morphologic traits.
- Biochemical traits (like isozymes, storage proteins).
- Molecular markers.

For agronomic and morphologic traits expressed in a continuous or ordinal scale one can quantify the difference between parents by calculating the Euclidean distance or the generalized distance. For biochemical and molecular marker data one may use the following measure for genetic similarity of genotypes *i* and *j*:

$$gs_{if} = \frac{2N_{ij}}{N_i + N_j}$$

Where

$$N_{ij} = Number\ of\ bands\ present\ in\ both\ i\ and\ j$$

$$N_i = Number\ of\ bands\ present\ in\ i$$

$$N_j = Number\ of\ bands\ present\ in\ j$$

Transgression may occur at a large genetic distance between potential parents. The greater the distance (up to a certain limit), the larger the number of segregating loci and the larger the probability of transgression.

- Information about *the performance as a parent of the pursued genotype(s)*. Such information is obtained from earlier breeding cycles or earlier test crosses (for example a diallel cross yielding information about general combining ability and about specific combing ability).
- Information about *the performance of early generation progenies* from crosses involving the potential parents. From these one can estimate the mean and the variance as expected to apply to later generations.

Sources 1 and 2 deal with qualitative traits, such as growth habit of barley lines, *viz. erectoides* versus *nutans*. Sources 3–5 deal with information about quantitative traits. Parents are crossed in such a way that weaknesses of one parent are compensated for by the other parent.

Jensen reviewed the topic of choosing parents extensively. Indeed, the association between genetic distance and probability of transgression has often been studied. A number of scientists advocated the crossing of parents with a low genetic similarity, but experimental evidence supporting this advice is scarce.

Crossing of divergent lines often yields populations with a low mean performance due to one of the parents involved. Linkage groups of favourable genes are broken at meiosis of the heterozygous plants. Such groups are difficult to recover in later generations. Brown and Caligari (1989) studied cross prediction based on evaluation of parental genotypes, or their offspring obtained after selfing. Thus mid-parent phenotypic values, *i.e.*

$$\frac{1}{2}(\overline{P}_{P1} + \overline{P}_{P2})$$

and mid – line phenotypic values, i.e.

$$\frac{1}{2}(\overline{P}_{L(P1)} + \overline{P}_{L(P2)})$$

In Section, it was shown that the latter two procedures may be expected to be reliable for traits where dominance does not play a role in the genetic control.

It has to be emphasized that information sources 4 and 5 do, in fact, not provide information with regard to crosses still to be made. They merely indicate which already existing segregating populations are most promising.

With regard to that topic, crosses, in fact segregating populations, may be ranked according to some criterion. In a self-fertilizing crop crosses may, for instance, be ranked according to

- Their ability to give rise to entries with a genotypic value exceeding some minimum, say *Gmin*. This may involve ranking of crosses with regard to P($G > Gmin$), *i.e.* the probability that the genotypic value of some obtained genotype exceeds *Gmin*. The probabilities are then predicted on the basis of estimates of m and $\sum_i a_i^2$
- The observed proportion of (F3) lines with a mean phenotypic value exceeding *Gmin*.

Reliability of the prediction of the performance of the progenies to be obtained when crossing parents is, of course, very desirable. Genotype by environment interaction is, of course, a disturbing phenomenon. If such interaction occurs, predictions on the basis of data collected in a certain macro-environment (year and/or location) will be of little value for other macro-environments. Furthermore the reliability of cross prediction methods is questionable in as far as the estimators of the statistical parameters are biased and/or inaccurate. In the case of a normal probability distribution of the genotypic values, *i.e.*

$$\underline{G} = N(E\underline{G}, \sigma_g^2),$$

one can predict $P(G > Gmin)$ on the basis of estimates of EG and $\sigma^2 g$. This is elaborated for plant material with identical reproduction and for self-fertilizing crops.

Cross prediction with regard to several traits deserves attention because selection is rarely focussed on only a single trait. The probability that an inbred line has a satisfactory genotypic value for two or more traits simultaneously cannot be calculated as the product of the probabilities for the separate traits, unless the traits are not correlated. Multivariate cross prediction procedures require, in addition to knowledge of m and of $\sum_i a_i^2$ for each character, also knowledge of the genetic correlation coefficient, ρg, between each pair of characters. Powell *et al.* present an application of multivariate cross prediction methods.

THE CONCEPT OF COMBINING ABILITY AS APPLIED TO PURE LINES

The genetic quality of a genotype appears often poorly from the phenotype of the plant(s) representing the genotype, especially when the genotype is represented by only a single or a few plants. An alternative way of assessing the genetic quality of the genotype is by means of evaluation of progeny obtained from it. Indeed, in cross-fertilizing crops the application of selection based on progeny testing, *i.e.* selection for breeding value, is quite common. Candidate genotypes, representing some genetically heterogeneous population, are then pollinated by a tester population producing pollen with a diverse haplotypic composition. Candidate genotypes yielding the best progenies are selected.

With regard to sets of pure lines something similar may be applied. The genetic quality of a pure line is then assessed on the basis of the progeny obtained by crossing the line with a tester population (in the present case consisting of a set of pure lines). This procedure may be applied to a selffertilizing crop but also to a cross-fertilizing crop. The latter situation applies when testing pure lines with the goal to develop a hybrid variety. Candidate genotypes producing the best performing offspring are said to have the highest combining ability. The crossing design of the lines to be assessed may consist of a diallel cross, sometimes indicated as: a diallel set of crosses. In this case all N pure lines are crossed in pairwise combinations. The diallel cross is said to be complete if each line is crossed with all other lines. This will yield *N*2 progenies, *viz.* *N* S1-lines due to selfing, and *N*2-*N* FS-families due to pairwise crosses. If selfing is omitted and reciprocal crosses are not made only 1 2*N*(*N*-1)

FS-families will be obtained.

In this book it is assumed that the *N* candidate genotypes are pure lines. They may be designated as P1, P2, ..., P*N*. The progenies may be coded as F*ij*, where

- *i* refers to maternal parent P*i*; with *i* = 1, ..., *N*
- *j* refers to paternal parent P*j*; with *j* = 1, ..., *N*

Each progeny may be represented by a single plant or by a number of plants that are either cultivated as individually randomized plants or as *J* plots each containing *K* plants. The quantitative genetic interpretation of the observation characterizing the single cross hybrid progeny F*ij* may thus range from 'the phenotypic value of a single plant representing the hybrid' to 'a precise estimate of the genotypic value of the hybrid'. For this reason the observation will be designated by the general symbol *xij*. Table presents a summary of the observations derived from all progenies resulting from a complete diallel cross.

Table: The observation xij characterizing progeny Fij obtained from a complete diallel cross involving pure lines P1,...,PN;i,j = 1,...N. The margins of the table provide for each maternal parent as well as for paternal parent the mean progeny performance

	Paternal parent					
	P_1	.....	P_j	.....	P_N	
P_1	x_{11}	.....	x_{1j}	.....	x_{1N}	$\bar{x}_{1.}$
.	.		.		.	.
.	.		.		.	.
P_i	x_{i1}	.....	x_{ij}	.....	x_{iN}	$\bar{x}_{i.}$
.	.		.		.	.
.	.		.		.	.
.	.		.		.	.
P_N	x_{N1}	.....	x_{Nj}	.....	x_{NN}	$\bar{x}_{N.}$
	$\bar{x}_{.1}$	.....	$\bar{x}_{.j}$	.....	$\bar{x}_{.N}$	$\bar{x}_{..}$

The set of progenies occurring in row *i*, *i.e.* $\{F_{i1}, ..., F_{iN}\}$, or the set of progenies occurring in column *j*, *i.e.* $\{F_{1j}, ..., F_{Nj}\}$, forms an HS-family, which may be designated by $F_{i}.$ and *F.j*, respectively. A row as well as a column comprises the observations from all progenies descending from the same maternal parent or the same paternal parent, respectively. The average across row *i*, say $\bar{x}_{i}.$, or across column *j*, say $\bar{x}_j$, represents the mean across the single cross hybrids constituting HS-family $F_{i}.$ or F_{j}, respectively. If the total number of 1 2*N*(*N*-1) progenies is unmanageably large, or if the breeder fails to produce all of them, for instance due to asynchronous flowering, a partial diallel cross (or incomplete diallel cross) may be made.This partial diallel cross may produce progenies according to a structured scheme, such as used for a balanced incomplete block design or an a-design, Example, or it may produce progenies according to an unstructured ('wild') crossing design. In the former case the maternal parents play the role of the treatments and the paternal parents the role of the incomplete blocks. Care must be taken for a wild crossing design that it is a connected design.In this book two reasons for making a diallel cross are elaborated 1. Prediction of the performance of a TC- or a DC-hybrid variety of a crossfertilizing crop. This application plays an important role in practical plant breeding aiming at the development of a hybrid variety.

- Determination of the general combining ability of a pure line and/or the specific combining ability of a pair of pure lines. This application occurs rather frequently at research stations, possibly in the framework of the development of a new variety.

GENERAL AND SPECIFIC COMBINING ABILITY

It is of interest to know whether or not a pure line possesses a good general combining ability (*gca*), with regard to a tester population; or whether two pure lines have a good specific combining ability (*sca*) or not. (The precise definitions of these quantities are developed hereafter. It should thus be clear that the main interest when applying an analysis in terms of *gca* and *sca* is not in the progenies but in their parents. An analysis of a diallel cross in these terms is, indeed, a special way of progeny testing.When applying a diallel cross the tester population consists of the set of inbred lines involved in the diallel cross. For inbred line i the value obtained for

$$\bar{x}_{i.} - \bar{x}_{..}$$

where

$\bar{x}_{..}$ designates the overall mean progeny phenotypic value, may be considered as an estimate of its general combining ability. Thus the general combining ability of a pure line is indeed estimated from the performance of its offspring in comparison to the overall mean performance.

One may subtract from the expected genotypic value, calculated across all progenies descending from pure line i, the expected genotypic value calculated across all progenies. The quantity obtained is similar to the breeding value of line i, except for the factor 2 occurring in Equation. The variance of the *gca* values is, consequently, similar to the variance of the breeding values. One should, nevertheless, be cautious. The concepts of additive genotypic value, breeding value, additive genotypic variance and variance of the breeding values are applied in the context of panmictic populations. Only in that situation Equation, *i.e.*

$$\sigma_a^2 = \text{var}(\underline{bv}),$$

Applies. In contrast the concepts of *gca* and *sca* apply to a different context, *viz.* to pure lines involved in a diallel cross.The concepts of *gca* and *sca* are also used in other contexts than diallel crosses, *e.g.* recurrent selection for gca, recurrent selection for sca, reciprocal recurrent selection. The concepts have, consequently, been defined in different ways.

Sprague and Tatum (1942), who introduced the terms gca and sca, used definitions different from those proposed by Griffing (1956). The approach of the latter, which is considered here, is similar to the one used for the statistical analysis of a two-way table. An analysis of the data resulting from a diallel cross in terms of *gca* and *sca* is thus primarily a statistical analysis. A two-way table may be analysed on the basis of a simple linear model

$$E\underline{x}_{ij} = \mu + \alpha_i + \beta_j + \gamma_{ij}$$

Such a model is also used for data obtained from a randomized complete block experiment such as used to compare the performances of a number of genotypes.

Griffing's parametrization of the genotypic value *Gij* of the single cross hybrid obtained by pollinating maternal parent *i* by paternal parent *j* is:

$$G_{ij} = \mu + gca_i + gca_j + sca_{ij}$$

In the case of a complete diallel cross yielding N_2 progenies the formulae for estimating the parameters μ, *gcai* and *scaij* in Equation are straight forward:

$$\mu = \overline{x}.. = \frac{\sum_{i=1}^{N}\sum_{j=1}^{N} x_{ij}}{N^2}$$

$$\hat{gca}_i = \frac{1}{2}(\overline{x}_i + \overline{x}_i) - \mu = \frac{\sum_{j=1}^{N} x_{ij} + \sum_{j=1}^{N} x_{ji}}{2N} - \hat{\mu}$$

$$\hat{sca}_{ij} = \frac{1}{2}\left(x_{ij} + x_{ji}\right) - \hat{gca}_i - \hat{gca}_j - \hat{\mu}$$

It is easily shown that the sum of the gca values is zero, namely

$$\sum_{t=1}^{N} \hat{gca}_i = \frac{1}{2}\sum_{t=1}^{N}(\overline{x}_i + \overline{x}_i) - N\hat{\mu} = \frac{\sum_{i=1}^{N}\sum_{j=1}^{N} x_{ij} + \sum_{j=1}^{N}\sum_{i=1}^{N} x_{ji}}{2N} - N\hat{\mu} = \frac{2N^2\hat{\mu}}{2N} - N\hat{\mu} = 0$$

This implies that the average gca value is bound to be zero. Likewise it is easily shown that for any line, for instance line i, the sum of the sca values is zero.

$$\sum_{j=1}^{N} s\hat{c}a_{ij} = \sum_{j=1}^{N}\left(\frac{1}{2}(x_{ij} + x_{ji}) - g\hat{c}a_i - g\hat{c}a_j - \hat{\mu}\right)$$
$$= \frac{1}{2}(x_i + x_i) - Ng\hat{c}a_i - N\hat{\mu}$$
$$= Ng\hat{c}a_i - Ng\hat{c}a_i = 0$$

Griffing (1956) elaborated the appropriate statistical analysis of data characterizing the progenies evolving from four different designs of a diallel cross, *i.e*. data from

- The N_2 progenies obtained from a complete diallel cross.
- All parental pure lines plus all *FS*-familes, reciprocals excluded, *i.e.*, NS_1-lines and $\frac{1}{2}N(N-1)$ *FS*-familes.
- All FS-families, reciprocals included, *i.e.* $N(N-1)$ FS-families
- All FS-families, reciprocals excluded, *i.e.* 1 $2N(N-1)$ FS-families

Both the analysis of variance according to a linear model assuming fixed effects and the analysis according to a linear model assuming random effects were elaborated. According to the model assuming fixed effects, the parents involved in the evaluated progenies are the subjects of study, whereas with the model assuming random effects interest is primarily in the population of pure lines represented by the random sample consisting of the N parents whose progenies were evaluated.

Designs 2 and 4 do not allow estimation of reciprocal differences, which may, for instance, be due to maternal effects via plasmagenes.

In Section, it was said that the genetic quality of a genotype might appear from an evaluation of its progeny. In the present section attention is focussed on progeny obtained from a diallel cross. An alternative for such progeny is the progeny obtained by selfing. Indeed, whenever a candidate has a valuable genotype its genetic value will appear from the quality of its offspring. The performance of offspring obtained by selfing is not at all affected by the tester genotype. Deleterious recessive genes hiding in the candidate genotype to be tested will clearly be exposed in the line obtained by selfing the candidate. For this reason, the authors are of the opinion that progeny testing of candidate genotypes by means of progenies obtained from selfing is a good alternative for progeny testing using progenies obtained from a diallel cross: it saves a lot of efforts (less crossing work, fewer progenies to be evaluated) and absence of disturbing tester effects (but possibly disturbing inbreeding effects due to the selfing; selfing might even be impossible due to self-incompatibility).

In Section, it was said that the genetic quality of a pure line can be assessed from the progenies resulting from a diallel cross in a way similar to the assessment of the breeding value of an open pollinating candidate. Indeed, an analysis of the data resulting from a diallel cross in terms of *gca* and *sca* is primarily a statistical analysis. It is, however, interesting to compare the pure line quantities *gca* and *sca* with the open pollinating candidate quantities breeding (*bv*) value and dominance deviation (δ). For this reason the quantitative genetic interpretation of the concepts *gca* and *sca* is developed (better than the rough quantitative genetic interpretation of *sca is* given.

The concept of breeding value applies to segregating populations of crossfertilizing crops; the concept of general or specific combining ability applies to sets of pure lines. There is, nevertheless, a rather close relationship between these concepts. In the absence of epistasis the expressions for *gca* and *sca* for a polygenic trait consist of the sum, across the involved loci, of the contributions due to individual loci. This requires the presence of linkage equilibrium when dealing with expressions for the variances of *gca* or *sca*.. The expressions of interest are thus derived from the expressions for locus *B-b*, affecting quantitative variation in a trait of an open pollinating population from which pure lines have been extracted. The relevant genotypic compositions are then

		Genotype		
		bb	*Bb*	*BB*
f:	In a panmictic population (RM):	q^2	$2pq$	p^2
	In a set of pure lines (L):	q	0	p

The expected genotypic values are

$$\mathrm{E}\underline{\mathcal{G}}_{\mathrm{RM}} = m + (p-q)a + 2pqd$$

$$\mathrm{E}\underline{\mathcal{G}}_{\mathrm{L}} = m + (p-q)a$$

A diallel cross yields FS-families. The genotypic composition of the aggregate of all FS-families is equal to the genotypic composition of the panmictic population. Thus EGFS = EGRM.

The genotypic composition of the HS-family obtained from a line with genotype *bb*, *i.e.* the set of all FS-families obtained from that line, is

	Genotype		
	bb	*Bb*	*BB*
f	*q*	*p*	0

The genotypic composition of the *HS*-family obtained from a line with genotype *BB* is

	Genotype		
	bb	*Bb*	*BB*
f	0	*p*	*p*

The general combining abilities of genotypes *bb* and *BB* may be designated by gca_0 and gca_2. respectively. They are equal to $EG_{HS} - EG_{RM}$. Thus

$$gca_0 = g(m-a) + p(m+d) - [m + (p-q)a + 2pqd] = pd - pa - 2pqd$$

$$= -p(a - d + 2qd) = -p[a - (1-2q)d] = -p[a - (p-q)d] = -p\alpha$$

It can likewise be shown that

$$gca_2 - q(m + d) + p(m + a) - [m + (p - q)a + 2pqd] - q\alpha$$

Comparison of the above results with Table show very simple relations between the above *gca* values and the *bv* values of the (homozygous) genotypes:

$$\underline{gca} = \frac{1}{2}\underline{bv} = \frac{1}{2}(\underline{\gamma} - E\underline{G})$$

$$\underline{bv} = 2\underline{gca}$$

The expected *gca* value, calculated across all homozygous genotypes, is easily obtained from the genotypic composition of the pure lines schema:

	Genotype	
	bb	*BB*
f	q	p
gca	$-p\alpha$	qa

Thus

$$E\underline{gca} = q(-p\alpha) + p(q\alpha) = 0$$

Furthermore

$$\text{var}(\underline{gca}) = E(\underline{gca})^2 - [E(\underline{gca})^2 = qp^2\alpha^2 + pq^2\alpha^2 = pq\alpha^2 = \frac{1}{2}\sigma_a^2$$

The results expressed by Equations may not be derived, via Equation, from E*bí* and var(*bí*) as the latter quantities apply to panmictic populations. Equation would, for instance, yield: var(*gca*) = 1/ 4 var(*bí*) = 1/ 4σa^2.

In the scheme below, the margins provide the relative frequencies of the maternal and paternal pure lines involved in the diallel cross (and their genotypes); the central part provides the relative frequencies of the various FS-families resulting from the diallel cross (and their genotypic compositions):

	$q(bb)$	$p(BB)$
$q(bb)$	$q^2(1,0,0)$	$pq(0,1,0)$
$p(BB)$	$pq(0,1,0)$	$p^2(0,0,1)$

The genotypic value of (genetically uniform!) FS-families with genotypic composition (1,0,0) is $m-a = G0$. It is $m+d = G1$ for FS-families with genotypic composition (0,1,0) and $m+a = G2$ for FS-families with genotypic composition (0,0,1).

The specific combining ability of genotypes *bb* and *bb*, of genotypes *bb* and *BB*, and of genotypes *BB* and *BB* are now designated by *sca*00, *sca*02 and *sca*22, respectively. According to Equation, they are equal to ,

$$sca_{ij} = G_{ij} - \mu - gca_i - gca_j,$$

$$\underline{G}_{FSij} - \mu - gca_{pi} - gca_{pj}$$

According to Equation the dominance deviation of a genotype belonging to a panmictic population is equal to the difference between its genotypic value and its additive genotypic value, where the additive genotypic value is equal to $\mu + bv$. Thus

$$\delta = G - \gamma = G - \mu - bv$$

This implies

$$sca_{00} = G_0 - \mu - 2gca_0 = G_0 - \mu - bv_0 = \delta_0$$

$$sca_{02} = G_1 - \mu - \frac{1}{2}bv_0 - \frac{1}{2}bv_2 = G_1 - \mu - bv_1 = \delta_1$$

$$sca_{22} = G_2 - \mu - 2gca_2 = G_2 - \mu - bv_2 = \delta_2$$

The *sca* value of a pair of homozygous genotypes appears thus to be equal to the dominance deviation of the corresponding F1 genotype. Alternatively, the other way around – the dominance deviation of a genotype is equal to the *sca* value of its homozygous parents.

The variance of the *sca* values of pairs of lines is calculated from the probability distribution of the various pairs of lines and their *sca* values, *i.e.*

Pair of lines

	(bb,bb)	(bb,BB)	(BB,BB)
f	q^2	$2pq$	q^2
sca	δ_0	δ_1	δ_2

This means that

$$E\underline{sca} = E\underline{\delta} = 0$$

and

$$\text{var}(\underline{sca}) = \text{var}(\underline{\delta}) = \sigma d^2$$

Furthermore Equation implies that the variance of the genotypic values of the progenies obtained from the complete diallel cross is equal to

$$\text{var}(\underline{G}) = \text{var}(\underline{gca}_M) + \text{var}(\underline{gca}_p) + \text{var}(\underline{sca}) = \sigma_d^2 + \sigma_d^2$$

where M and P refer to the maternal and paternal lines, respectively.

In conclusion, the quantitative genetic interpretation of the statistical quantities *gca* and *sca* is in terms of breeding values, additive genotypic values and dominance deviations. In the absence of overdominance one may state that the *gca* value of a line will be high if it has, for many loci, the homozygous genotype *BB*, giving rise to a good performance. Then lines with a good *gca* will tend to have a good performance *per se*. Improvement of *gca* can then simply be pursued by elimination of undesired recessive alleles, *e.g.* by line selection. This means that a diallel cross, made with the single goal to evaluate *gca* values, is a waste.

The observation that a cross between certain inbred line yields an unexpectedly good performing offspring is, nevertheless, of direct significance when developing a SC-hybrid variety. The *gca* of a pure line and the *sca* of a pair of pure lines depend on the set of pure lines used as a tester. Thus, estimates of *gca* and *sca* derived from a particular diallel cross do not apply to other sets of pure lines. In this sense estimation of *gca* and *sca* is of minor significance. For an incomplete diallel cross one may, however, predict the genotypic value G_{ij} of any FS-family F_{ij} which was not actually generated, by

$$\widehat{G}_{ij} = \bar{x}.. + \widehat{gca}_i + \widehat{gca}_j$$

If the *sca* effects, *i.e.* the dominance deviations, are of minor importance, this approach may save considerable efforts otherwise to be dedicated to crossing and testing. It is speculated that this possibility of predicting progeny performance is insufficiently exploited.

The timing of the estimation of the combining ability of inbred lines deserves attention. In maize breeding it is still current procedure to develop pure lines by selfing for 5-7 generations. Until this stage only some visual selection is applied, but – because it has often been observed that the performances of inbred lines do not predict precisely enough the performance of the SC-hybrid to be obtained from these lines – the selection is useless with regard to the performances of the hybrids to be made. Thereafter the combining abilities of the more or less pure lines are determined.

Effort-saving shortcuts are, of course, attractive. Consequently, it is of interest to check how well the performances of progenies obtained by crossing 'young' inbred lines predict the performances of the hybrids obtained by crossing pure lines tracing back to these young lines. The limits of the potentials of the inbred lines derived from some S0 plant are *a priori* determined by the genotype of the S0-plant. Thus a reliable procedure for early assessment of the potentials of lines under development would be of great value. It would allow breeders to devote more efforts to selection among lines from S0 plants that appeared to be promising. Jenkins (1935) came to the conclusion that the 'genetic values' of inbred lines, evaluated by testing progenies obtained from top-crosses, are determined early in the inbreeding process. This led to the evaluation procedure called early testing. It was aimed at the identification of young lines deserving further development.

PLANT MATERIAL WITH IDENTICAL REPRODUCTION

This section gives attention to the prediction of the ranking of crosses dealing with plant material with identical reproduction, *e.g.* clones, pure lines (especially DH-lines). The conditions required for a reliable prediction of the probability that the genotypic value of some genotype exceeds some minimum, *i.e.* $P(G > Gmin)$, are

- A normal distribution of the genotypic values
- Absence of genotype × environment interactions

When estimating EG by p and var(G) on the basis of a completely randomized experiment or randomized (complete) blocks, one may predict $P(G > Gmin)$ by:

This probability can be read from a table presenting values of the standard normal distribution. The probability can be predicted for each of a number of families ('crosses') and this allows ranking of the crosses. The coefficient of correlation between predicted rank and actual rank indicates the reliability of the prediction.

Self-Fertilizing Plant Material

If the genotypic values of the homozygous genotypes in an F– population of a self-fertilizing crop have a normal distribution, the probability distribution of G is completely specified by EG and σ^2 g. Under the conditions specified below, one may predict these parameters from data collected from the parents and from a random sample of F3 lines.

Then one may predict the probability that the genotypic value of an F– plant exceeds *Gmin*.

The conditions required for a reliable prediction are the following:

- A normal distribution of the genotypic values.
- Absence of epistasis.
- Absence of linkage.
- Absence of genotype × environment interactions.

If condition 1 applies the probability distribution of the genotypic values of the plants in population F∞ is given by

$$\underline{G} = N(m, \mathrm{var}(\underline{G}_{F\infty})$$

Condition 2 is required to estimate parameter m by means of Equation

$$\widehat{m} = \frac{1}{2}(\overline{P}_{P1} + \overline{P}_{P2})$$

If conditions 2 and 3 are satisfied, var(GF∞) is equal to. $\sum_i a_i^2$

A biased but relatively accurate estimate of this quantity is 2v^ar(GLF3). The probability distribution of F– can thus be predicted. An interesting application, *i.e.* prediction of $P(G > Gmin)$, requires condition 4. If the condition applies, the probability that some F– plant to be obtained in the future has a genotypic value exceeding *Gmin*, is predicted by:

$$P\left(\frac{\underline{G}-\widehat{m}}{\sqrt{\widehat{var}(\underline{G}_{F\infty})}} > \frac{G_{\min}-\widehat{m}}{\sqrt{\widehat{var}(\underline{G}_{F\infty})}}\right) = P\left(\underline{x} > \frac{G_{\min}-\widehat{m}}{\widehat{\sigma}g}\right) = 1-\phi\left(\frac{G_{\min}-\widehat{m}}{\widehat{\sigma}g}\right)$$

Calculation of this probability may be rewarding. When for two segregating populations the means $m1$ and $m2$ and the genetic variances v^ar1(*G*F∞) and v^ar2(*G*F∞) differ such, that $m1 > m2$ and v^ar1(*G*F∞) < v^ar2(*G*F∞), then it is of interest to calculate *P*(*G* > *Gmin*) for each population.

In addition to the foregoing, one may perhaps wish to predict the genotypic values of the two extreme homozygous genotypes (Jinks and Perkins, 1972). These values are

$$m-\sum_i a_i \text{ and } m+\sum_i a_i$$

Prediction of these values requires estimates of m and $\sum_i a_i$. The latter quantity may be estimated when assuming a constant degree of dominance across all relevant loci, *i.e.*:

$$\frac{d_i}{a_i}=c$$

Then one may derive

$$\sum_i d_i.\sqrt{\frac{\sum_i a_i^2}{\sum_i d_i^2}} = \sum d_i.\sqrt{\frac{\sum a_i^2}{c^2\sum a_i^2}} = \sum d_i.\frac{1}{c} = \sum d_i.\frac{\sum a_i}{\sum d_i.} = \sum_i a_i$$

According to Table the quantity $\sum_i d_i$ may be estimated by

$$\widehat{G}_{F1}-\widehat{m}$$

The quantity

$$\sum_i a_i^2$$

is estimated as

$$2\widehat{var}(\underline{G}_{LF3})$$

and

$$\sum_i d_i^2$$

can, for instance, be estimated on the basis of Equations. The reliability of this approach for estimating is questionable. In the case of presence of one or more loci with additive effects, for instance, it yields a false result.

PREDICTION OF THE RESPONSE TO SELECTION

When dealing with selection with regard to quantitative variation the concepts of selection differential, designated by S, and response to selection, designated by R, play a central role. These concepts, Figure, are defined as follows:

$$S := E\underline{p}_{s,t} - E\underline{p}_t$$

where
$$R := E\underline{p}_{t+1} - E\underline{p}_t$$

- $E\underline{p}_{s,t}$ designates the expected phenotypic value of the candidates (plants, clones, families or lines) in generation t of the considered population with a phenotypic value greater than the phenotypic value minimally required for selection (p_{min}). $E\underline{p}_{s,t}$ designates thus the expected phenotypic value of the selected candidates.
- $E\underline{p}_t$ designates the expected phenotypic value calculated across all candidates belonging to generation t of the population subjected to selection.
- $E\underline{p}_{t+1}$ designates the expected phenotypic value calculated across the offspring of the selected candidates.

The selection differential (S) in generation t and the response to the selection (R) are indicated. The shaded area represents the probability that a candidate has a phenotypic value larger than the minimally required phenotypic value (p_{min}) it was derived that

$$E\underline{P} = E\underline{G}$$

The response to selection is now considered for three situations:

- The hypothetical case of absence of environmental deviations, as well as absence of dominance and epistasis.
- Absence of environmental deviations, presence of dominance and/or epistasis.
- Presence of environmental deviations, dominance and/or epistasis.

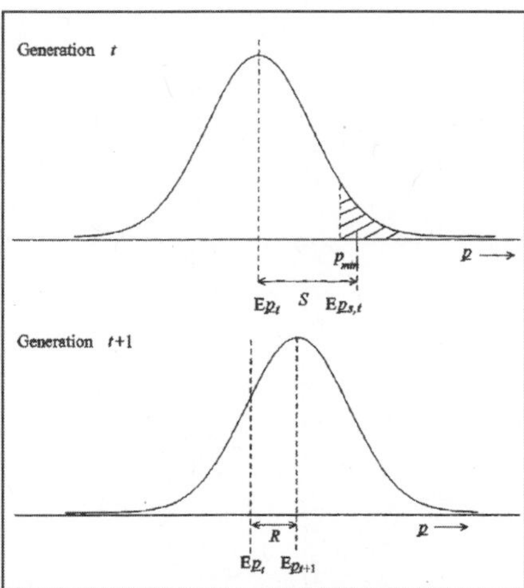

Fig. The density function for the phenotypic value p in generation t and in generation $t+1$, obtained by selecting in generation t all candidates with a phenotypic value greater than p_{min}.

Absence of Environmental Deviations, Dominance and Epistasis

In the absence of environmental deviations, dominance and epistasis, both the genotypic value and the phenotypic value of a candidate can be described by a linear combination of the parameters $a_1, ..., a_K$ defined in Section. Selection of candidates with the highest possible phenotypic value implies selection of candidates with genotype $B_1B_1 ... B_KB_K$ and with genotypic value. The offspring of these candidates will have the same phenotypic and genotypic value as their parents. This applies to self-fertilizing crops as well as cross-fertilizing crops, when the selection occurs before pollen distribution. Under the described conditions R will be equal to S.

Absence of Environmental Deviations, Presence of Dominance and/or Epistasis

In the case of absence of environmental deviations but presence of dominance and/or epistasis, selected candidates, with the same highest possible phenotypic value, may have a homozygous or a heterozygous genotype. Then the offspring of the selected candidates are expected to comprise plants with genotype *bb* for one or more loci, giving rise to an inferior phenotypic value compared to that of the selected candidates.

In the case of complete dominance, for instance, candidates with the highest possible phenotypic value for a trait controlled by loci $B_1 - b_1$ and $B_2 - b_2$ will have genotype $B_1{\cdot}B_2$. Selection of such candidates will yield offspring including plants with genotype $b_1b_1b_2b_2$, $b_1b_1B_2$ or $B_1{\cdot}b_2b_2$, having an inferior genotypic and phenotypic value. Under these conditions R will be less than S.

Presence of Environmental Deviations, Dominance and/or Epistasis

In actual situations environmental deviations, dominance and epistasis should be expected to be present. Among the selected candidates their phenotypic values will tend to be (much) higher than their genotypic values.

Furthermore, except in the case of identical reproduction, the genotypic composition of the selected candidates will deviate from that of their offspring. Under these conditions R will be (much) smaller than S. Selected maternal plants coincide with the selected paternal plants in the case of self-fertilizing crops, as well as in case of hermaphroditic cross-fertilizing crops if the selection is applied before pollen distribution. In other situations, the set of selected maternal parents providing the eggs differs from the set of selected paternal parents providing the pollen. Then one should determine S_f for the candidates selected as maternal parents and S_m for the candidates selected as paternal parents. Because both sexes contribute equal numbers of gametes to generate the next generation we may write

$$S = \frac{1}{2}(S_f + S_m)$$

Equation does not only apply at selection in dioecious crops, but also when selecting in hermaphroditic cross-fertilizing crops when the selection is done after pollen distribution. In the latter case there is no selection with regard to paternal parents. This implies $S_m = 0$ and consequently $S = 1\ 2S_f$

Actual situations tend to be more complicated. Consider selection before pollen distribution with regard to some trait X. In the case of an association between the expression for trait X and the expression for trait Y, the selection differential for X implies a correlated selection differential with regard to Y, say *CS*. Thus

$$CS_Y := E\underline{P}_{Ys,t} - E\underline{P}_{Y,t}$$

where

- E*p* *Ys,t* designates the expected phenotypic value with regard to trait Y of the candidates selected in generation *t* because their phenotypic value with regard to trait X being greater than minimally phenotypic value (*pXmin*).
- E*p* *t* designates the expected phenotypic value with regard to trait Y calculated across all candidates belonging to generation *t* of the population subjected to selection with regard to trait X. When considering a linear relationship between the phenotypic values for traits X and Y, the coefficient of regression of

$$\beta_{PY,PX} = \frac{\text{cov}(\underline{P}_Y, \underline{P}_X)}{\text{var}(\underline{P}_X)}$$

$$CS_Y = \beta_{PY,PX} S_X$$

The indirect selection for trait Y, via trait X, may be followed, after pollen distribution, by direct selection for Y. The effective selection differential for Y comprises then a correlated selection differential.If the considered trait has a normal distribution, E*p* *s,t, i.e.* the expected phenotypic value of those candidates with a phenotypic value larger than the value minimally required for selection, may be calculated prior to the actual selection. This will now be elaborated.

A normal distribution of the phenotypic values for some trait is often designated by

$$\underline{p} = N(\mu, \sigma^2)$$

Where

$$\mu = E\underline{P}, \textit{ and}$$

$$\sigma^2 = \text{var}(\underline{P})$$

Standardization, *i.e.* the transformation of *p* into *z* according to

$$\frac{\underline{p} - \mu}{\sigma} = \underline{z}$$

implies that *z* has a standard normal distribution characterized by

$$\mu_z = 0 \textit{ and}$$
$$\sigma_z = 1$$
$$\underline{z} = N(0,1)$$

Selection of candidates with a phenotypic value exceeding the phenotypic value minimally required for selection (*pmin*) is called truncation selection. Selection of superior performing candidates up to a proportion *v* implies applying a value for *pmin* such, that

$$v = P(\underline{P} > P_{\min})$$

Standardization of Pmin yields the standardized minimum phenotypic value $z_{\min}$:

$$v = P(\underline{P} > P_{\min}) = P(\underline{z} > z_{\min}) = \int_{z_{\min}}^{\infty} f(z).dz$$

where $$z_{\min} = \frac{P_{\min} - \mu}{\sigma}$$

Thus $$f(z) = \frac{1}{\sqrt{2\pi^{e}}} - \frac{1}{2}z^{2}$$

is the density function of the standard normal random variate *z*.

In Fig. the shaded area corresponds with *v*. Most statistical handbooks contain for the standard normal random variate z a table presenting *zmin* such P(*z* > *zmin*) is equal to some specified value *v*. Then one can calculate *pmin* according to

$$P_{\min} = \mu + \sigma z_{\min}$$

To measure the selection differential in a scale-independent yardstick, a parameter, called selection intensity and designated by the symbol *i*, has been defined:

$$i = \frac{S}{\sigma}$$

There is a simple relationship between the proportion of selected candidates (*v*) and *i* if the phenotypic values of the considered trait follow a normal distribution, namely

$$i = \frac{f(z_{\min})}{v}$$

where *f*(*zmin*) represents the value at *z* = *zmin* of the density function of the standard normal random variate *z*.

Also the measurement of the response to selection (*R*) deserves closer consideration. It requires determination of E*p* in the two successive generations *t* and *t* + 1. To exclude an effect of different growing conditions these two generations should preferably be grown in the same growing season. This is possible by

- Testing simultaneously plant material representing generation *t* + 1 (say population P_{t+1}), obtained by harvesting candidates selected in generation t, and – from remnant seed – plant material representing generation t (say population Pt)

- Testing simultaneously plant material representing generation $t + 1$, obtained by harvesting candidates selected in generation t (population P_{t+1}), and plant material, also representing generation $t + 1$, obtained by harvesting in generation t random candidates (population P_{t+1})

Simultaneous Testing of Populations P_{t+1} and P_t

Measurement of R by simultaneous testing of populations P_{t+1} and Pt will be biased if these populations differ due to other causes than the selection. Such differences may be due to

- The fact that the remnant seed is older and has, consequently, lost viability.
- The remnant seed representing Pt was produced under conditions deviating from the conditions prevailing when producing the seed representing P_{t+1}.
- A difference in the genotypic compositions of P_{t+1} and P_t which is not due to the selection. This is to be expected when dealing with self-fertilizing crops: P_{t+1} tends to contain a reduced frequency of heterozygous plants in comparison to Pt.

When testing populations P_{t+1} and Pt simultaneously, no allowance is made for the possible quantitative genetic effect of the reduction of heterozygosity occurring in self-fertilizing crops.

Simultaneous Testing of Populations P_{t+1} and P_{t+1}

The causes for the bias mentioned above do not apply to simultaneous testing of populations P_{t+1} and P_t. Furthermore, this method allows – for crossfertilizing crops – estimation of the coefficient of regression of the phenotypic value of offspring on parental phenotypic value.

Such an estimate may be interpreted in terms of the narrow sense heritability. One should realize that R as defined by Equation does not represent

A lasting response to selection if For self-fertilizing crops populations after generation $t + 1$, obtained in the absence of selection, will – due to the ongoing reduction of the frequency of heterozygous plants – tend to have an expected genotypic value deviating from $Ep\ t+1 = Ep\ t + R$.

The same applies to selection after pollen distribution in cross-fertilizing crops: population P_{t+1} results then from a bulk cross and will, consequently, contain an excess of heterozygous plants compared to population P_{t+2} obtained – in the absence of selection – from population P_{t+1}.

In the case of selection before pollen distribution, population P_{t+1} is in Hardy–Weinberg equilibrium and P_{t+1} and P_{t+2} will then, in the absence of epistasis, have the same expected genotypic value.

A procedure to predict R is, of course, of great interest to breeders, because such prediction may be used as a basis for a decision with regard to further breeding efforts dedicated to the plant material in question. As the prediction is based on linear regression theory, a few important aspects of that theory are reminded. In the case of linear regression of y on x the y-value for some x-value is predicted by

$$\hat{y} = \alpha + \beta x,$$

where

$$\beta = \frac{\text{cov}(\underline{x}, \underline{y})}{\text{var}(\underline{x})} = \frac{E(\underline{x}.\underline{y}) - (E\underline{x}).(E\underline{y})}{E\underline{x}^2 - (E\underline{x})^2}$$

and because of

$$E\underline{y} = \alpha + \beta.E\underline{x}$$

the intercept α is equal to

$$\alpha = E\underline{y} - \beta.E\underline{x}$$

Thus

$$\hat{y} = (E\underline{y} - \beta.E\underline{x}) + \beta x = E\underline{y} + \beta(x - E\underline{x})$$

implying

$$\hat{y} = E\underline{y} = \beta(x - E\underline{x})$$

$$E\underline{P}_{t+1} - E\underline{P}_t = \beta(E\underline{P}_{s,t} - E\underline{P}_t)$$

$$R = \beta S$$

It is common practice to substitute parameter β in Equation either by the wide or by the narrow sense heritability:

- In the case of identical reproduction, this applies when dealing with clones, pure lines and single-cross

$$\frac{\sigma g^2}{\sigma p^2}$$

Hybrids, β is substituted by the ratio, *i.e.* the heritability in the wide sense, commonly designated by *hw*

Thus

$$R = hw^2 S$$

In this situation the genotypes of the selected entries are preserved. Note presents the derivation of Equation.

- In the case of non-identical reproduction of the selected candidate plants of a cross-fertilizing crop b is substituted by, *i.e.* the heritability in narrow sense, commonly designated by *hn*

Thus.

$$R = h_n^2 S$$

The possible bias introduced with this substitution is taken for granted.

The coefficient of correlation of $\underline{G}$ and Pg,p, is equal to the square root of the heritability in the wide sense:

$$p_{g,p=h_w}$$

The coefficient of regression of $\underline{G}$ on P, i.e. β is equal to the heritability in the wide sense:

$$\beta = h_w^2$$

The coefficient of correlation of the additive genotypic value and is equal to the square root of the heritability in the narrow sense:

$$p_{\gamma,p} = h_n$$

Because $S = i\sigma$ can also be written as

$$R = \beta.i\sigma$$

Equation can thus be written as

$$R = h_w^2 i\sigma_p = ih_w\left(\frac{\sigma_g}{\sigma_p}\right)\sigma_p = ih_w\sigma g$$

When selecting, after pollen distribution, in a cross-fertilizing crop one can similarly write

$$R = \frac{1}{2}ih_n^2 i\sigma_p = \frac{1}{2}ih_n\left(\frac{\sigma_a}{\sigma_p}\right)\sigma_p = \frac{1}{2}ih_n\sigma a$$

Higher selection intensities occur at lower proportions of selected plants. One should thus be careful when using the terms 'selection intensity' and 'proportion selected candidates.'

Macro-Environmental Conditions Maximizing σg^2 or h^2

It can be said that a breeder should look for macro-environmental conditions such, that the heritability is high. This requires the macro-environment to be uniform, *i.e.* σe^2 is small, and the genetic contrasts to be large, *i.e.* σg^2 is large.

However, for different traits different sets of macro-environmental conditions may then be required. For example: selection for a high yield per plant may require a low plant density, but selection for a high yield per m^2 may require a high plant density.

For traits with a negligible genotype × environment interaction the selection may be done on the basis of testing in a single environment. Thus in order to select in oats for resistance against the crown rust disease, a number of oat genotypes may be inoculated in the laboratory with crown rust fungal spores.

This maximizes the heritability of the degree of susceptibility (differences in the susceptibility do not show up in the absence of the disease). Then (on the assumption that laboratory tests are reflected in field performance) all resistant oat genotypes are expected to be resistant under commercial growing conditions.

For traits with important $g \times e$ interaction, however, selection in the single macro-environment yielding maximum heritability may imply selection of genotypes that do not perform in a superior way in the target environment.

Macro-EnvironmentalConditionsIdentical to those of the Target Environment

The suggestion to select under macro-environmental conditions identical to those of the target environment is generally accepted as a good guideline. However, with regard to plant density this suggestion implies a problem: due to the intergenotypic competition occurring when selecting under the high plant density applied at commercial cultivation, candidates may be selected that perform disappointingly when grown *per se, i.e.* in the absence of intergenotypic competition. Intergenotypic competition is a phenomenon which does not show up in the target environment provided by farmers growing genetically uniform varieties. With regard to competition it is, in fact, impossible to apply selection under conditions identical to those of the target environment. This topic is further considered in Section.

Fasoulas and Tsaftaris suggested that breeders should provide favourable growing conditions when selecting. The latter seems to be supported by the results of the experiment mentioned in Example, but the example also supports the idea that selection should be done under macro-environmental conditions similar to those of the target environment. Example illustrates that selection aiming to increase grain yield under less-favourable conditions was the most effective when applied under the poor conditions of the target environment.

Macro-Environmental Conditions Characterized by Absence of Interplant Competition

The idea of avoiding interplant competition by applying a very low plant density is supported by the problem indicated in the former paragraph. Gotoh and Osanai (1959) and Fasoulas and Tsaftaris (1975) advocated application of selection at such a low plant density that interplant competition does not occur.

An objection against selecting at a very low plant density is its inefficiency if genotype $\times$ plant density interaction occurs. Thus some (*e.g.* Spitters) have defended the opinion that selection should be applied at the plant density of commercial cultivation. This, however, would generate the problem of intergenotypic competition, a problem not occurring at a very low plant density.

Four reasons for such a discrepancy are mentioned here:

- If linkage and/or epistasis occur, estimators for the heritability based on the assumption of their absence are biased.
- The estimators of the heritability have some inaccuracy.

- The macro-environmental conditions experienced by population P_t, the population subjected to selection, may differ from those experienced by population P_t+1, the population obtained from the selected candidates. This relates both to imposed conditions, such as plant density, and uncontrollable conditions, such as climatic conditions. The actual response, appearing from a comparison of populations P_t+1 and P*t*, is then to be regarded as a correlated response due to indirect selection P*t*. In this situation the result of deliberate selection is sometimes hardly better than the result of 'selection at random'.
- Because the phenotypic values for different quantitatively varying traits tend to be correlated, selection with regard to a certain trait implies indirect selection with regard to other, related traits. The correlated response to such indirect selection may turn out to be negative with regard to pursuing a certain ideotype.

The indirect selection for biomass of maize, via selection for plant volumes, for instance, gave rise to a population susceptible to lodging. In the long-lasting selection programme of maize described in Example, selection for oil content implied indirect selection with regard to many other traits. A correlated response to selection was observed for: grain yield, earliness, plant height, tillering, *etc.*

Notwithstanding the often observed discrepancy between the predicted and the actual response to selection, the relation $R = \beta S$ is for plant breeders one of the most useful results of quantitative genetic theory. Based on this relationship the concept of realized heritability, designated as *hr* 2, has been defined. It is calculated after having established the actual response to selection at some selection differential. When selecting among identical reproducing candidates, or when selecting before pollen distribution in a population of a cross-fertilizing crop the definition is

$$h_r^2 = \frac{R}{S}$$

Because R has already been established, the quantity $h_r{}^2$ can not be used to predict R. It indicates afterwards the efficiency of the applied selection procedure.

$$h_r^2 = \frac{2R}{S_f}$$

8

Polyploid Breeding

Cytological study of polyploid species shows why most of them behave in bteeding like the hybrid polyploid of the Primula kewensis type. Whether they are tetraploids, hexaploids or octoploids, their chromosomes usually unite in pairs at reduction. This rule is particularly strict in those plants which depend on seed fertility for their reproduction, *e.g.*, the cereals. In these the differentiation is sharper even than in Primula kewensis. It is more like that in the Raphanus-Brassica tetraploid, where no pairing takes place between corresponding chromosomes of the two species even in the diploid hybrid. They are, as it were, in two watertight compartments. The species behave on self-fertilisation almost exactly as though they were diploids. Yet, apart from their chromosome numbers, there are three ways in which these species show us that they are polyploids.

In the first place, a particular character in a polyploid may be given by the presence of any one of several different factors. These factors are probably in corresponding chromosomes of the different diploid ancestors of the polyploid. In the hexaploid bread wheat, Tritkum vulgare, Nilsson-Ehle found that some varieties may have three independent factors affecting the red colour of the grain.

In the hexaploid oats, Akerman found similar factors affecting the presence of green leaf-colour. Other characters are affected by-two factors. We do not know that these multiple factors are due to the plant being a polyploid, but since they are not usually found in diploid plants, it is very probable. The occurrence of two independent factors for black grain-colour has been turned to advantage in breeding oats in Sweden. Old black varieties of oats had a single factor which may be called BXBX for black colour. These varieties used to sport so as to give a hundred or so white grains in every bushel, owing to the corresponding chromosomes in the second set carrying the recessive factor b9 and occasional pairing taldng place between the B and b chromosomes. The white grains, of course, reduced the value of the oats by giving it unfairly an appearance of impurity. Akerman x therefore crossed two black varieties, and from the F2 a new type doubly black BXBXB2B2 was raised, which never sports to white seeds. Secondly, when diploid species are X-rayed their offspring change in every direction; polyploids do not vary anything like so much as their diploid relatives. While tetraploid species of oats and wheat did mutate a little under the same doses as gave big changes in the diploids, the hexaploid bread wheat and cultivated oats showed no change whatever. What is true of change induced by X-rays is probably equally true of natural change. The extra chromosome sets act as a buffer against the effects of changes in the chromosomes, so that in polyploids we can have very little idea of what is going on inside the chromosomes, as compared with diploids. The doubly black-seeded variety of oats is an example of this; the second pair of factors is there for safety and if mutation of Bx to bx occurs, we shall not notice it.

But we can occasionally see the results of other changes, which provide a third kind of evidence of polyploidy. Diploid pure lines sometimes breed true, and for generations show no variation to the closest scrutiny: such are known in the Triticum monococcum, of de Vilmorin, which has bred true for eighty years, and in Johanssen's pure lines of broad bean, Vicia Faba.

But the hexaploid cereals cannot be depended upon. Awnless cultivated oats of as pure race as can be got always throw a proportion—varying with the variety and particularly high in new varieties—of individuals having awns on their flowering glumes. And when these are selfed, their progeny (which will appear in cultivation unless oats for seed are severely rogued) are of quite a distinct and inferior type, indistinguishable, in fact, from the wild oat, Avena fatua. These are called "fatuoid" oats. They are characterised by having both the spikelets provided with a twisted, kneed awn, and each grain marked by a prominent articulation surrounded by a tuft of hairs.

The chromosome behaviour at reduction that the "mutants" are due to the same cause as the appearance of the mealy offspring of P. kswensis, which also resemble a related species, P. verticillata* In both species of Avena, A. fatua and A. sativa each chromosome type is represented by three pairs. Taking two of these pairs -we may consider that A. fatua has the constitution FF9FF9 while A. Sativa has the constitution FF9ff9 and regularly produces gametes Ff which reproduce it truly. Both, as a rule, breed true, but occasionally in A. sativa (particularly in new varieties which have been disturbed by hybridisation) F pairs with/, or the four chromosomes FFff form a ring at reduction. The result is the formation of germ-cells that are FF or ff instead of the normal Ff.

When these meet the normal germ-cell the result is the production of two new types—FFFf and Ffff. The first of these is our slightly awned type, which when selfed gives, together with the normal and its own type, the pure fatuoid, FFFF, in the proportions. The same result is given by the first cross between the wild oat and cultivated oat, so that we know in this case roughly what makes the difference between the two.This property of giving away its hybrid origin is probably very common in new polyploid species, for the bread wheats do just the same thing. They throw a sport called a "speltoid" in exactly the same way as the oats throw a fatuoid. Again, this form resembles a related species, Triticum Spelta, in the character of its spike and grain. Such behaviour shows that rogueing is a much more serious business with polyploids than with diploids. All the materials are there for them to produce a whole array of sports, and if they are left to themselves although they are continually self-fertilised they will none the less gradually become more and more adulterated by their own throw-backs, which will mainly be of a character unwanted in cultivation.

MICRO-ORGANISMS

The importance of cultivated micro-organisms in agriculture and medicine is rapidly increasing. Most of them are unicellulars, and only in a few cases (yeasts) is it possible to do any cross-breeding to enhance the potential variability. Winge, in Copenhagen, has done a great deal of work on yeasts, and has proved that controlled cross-breeding under microscopical control is quite possible. In most bacteria and uni-cellular algae, even in some protozoa, vegetative multiplication goes on for long series of generations, even if occasionally wholly new variability may be the result of what in higher organisms we should call cross-breeding. In the second place, true mutations may occur, and it is even possible that in this material, where the cytoplasm is of great importance compared to the nucleus (or where there may be no nucleus at all), quantitative shifts in the relative proportion between the substances present may be passed on from one generation to many subsequent ones ("Dauermodifikation "). Of course the possibility of adapting the material to our needs is due to the fact that the cultures are variable mixtures. In this respect "wild" cultures are always furnishing very valuable material for selection if we want novel qualities or if we want to improve existing qualities.

We must always bear in mind that almost always two sets of desiderata exist: the actual quality of the strain for its ultimate purpose, and the ease with which it is multiplied commercially. To give a striking example. It used to be necessary in the production of penicillin to grow the moulds on the surface of sheets of culture medium. After a while, strains were found that would grow when submerged in a liquid medium, like yeasts, and this made a continuous system of manufacturing possible, and materially reduced the cost of the methods, and so of the final product.

Once a valuable strain is obtained it is of great importance to keep it pure. The ideal method should be the isolation of one single spore, or one single cell. The usual method of the bacteriologists—that of isolating a single colony from a very dilute culture on a semi-solid plate—is also good enough in practice to bring down the potential variability. If we start new cultures continually from very small beginnings we can get round the difficulty of the unavoidable introduction of "wild" strains into our lots.

In vaccination with living microbes, such as in the treatment of young individuals with harmless strains of dreaded infections (contagious abortion in heifers, tuberculosis in human babies), it is evident that only bacteriological methods of isolation and reproduction are good enough. In the commercial production of antibiotics and yeast this is much less pressing. The same is true of nodule-bacteria in legumes.

The preparation of pure cultures for the production of Yoghourt and cheeses that must carry special qualities and aroma is certainly the province of the bacteriologist-specialist. The use of living cultures of micro-organisms in the fight against diseases is only in its infancy. In many cases where what is called "a mild infection" prepares the animals against "heavy" attacks (coccidiosis in poultry and in rabbits and fur-animals is a good example) veterinary science is bound to progress towards the use of strains of low aggressiveness. Some micro-organisms can be used to transform cheap material into foods of very high value. Yeasts are capable of making use of predigested wood pulp, and of producing feeding stuffs rich in many vitamins and proteins of very high food value. The solution of the green-food problem in animals and birds in close confinement can be solved by the plant breeder who develops suitable strains of unicellular algae (Chlorella and others).

The same methods of selection and of progeny testing can be applied to the culture and the breeding of micro-organisms that our breeders of domestic animals and of cultivated plants apply in their work. This is a very new field, and we cannot see where its development may lead us. But one thing is quite apparent, that it promises to become of enormous practical importance.

Drugs and Spices

Spices and drugs have the double advantage for the plant breeder that they have not been worked with, and that the product generally brings a very good price. Even plants that are habitually collected from the wild may very well repay cultivation; such plants as Denis, foxglove and valerian, for which there is a steady demand.

It will often pay to take such a plant into cultivation, to study its biology and try to produce it in favourable conditions, sometimes in a foreign country. Plants, compared with animals, have the advantage under conditions of cultivation, of staying where we put them; we can devote all our energy to improving their quality and productiveness.

Drugs and spices vary enormously in quality, and it is obvious that it must be the task of the plant breeder to improve both yield and quality. In this matter we are severely handicapped by the peculiarities of the trade. In every plant we must pass through a stage in which the "wild" product brings a much better price than the plantation product. This can be a matter of plain prejudice, in which the buyers know the looks and the smell of the wild stuff, and are suspicious when anything of a new kind is offered. Sometimes matters are reversed after a while (plantation quinine bark) when the trade begins to appreciate the content in an actual, measurable active sub- stance; but, even so, appearance counts for much (lichen and moss on cinchona bark and colour of the pods in vanilla). Breeding for high quality as expressed in chemical content will always pay in the end if we can find a market which is selling the active principle pure, or that is selling a standardized article (Denis and Digitalis).

A good example is the production of tobacco for the manufacture of nicotin. Tobacco species differ enormously in yield per acre: some T. rustica lines are remarkable yielders, both in regard to quantity and nicotin content. It is understood that some drug-plants cannot be compared by actual objective standards. They may have no actual value at all, and the growers of ginseng have no hope of ever getting as good a price for their product as for wild ginseng. Their only hope lies in selecting their cultivated plants and treating them during growth in such a way that they can be palmed off as wild. Nothing is as fickle as the trade in drugs and spices. The manufacture of synthetic vanillin or alizarin may wipe out the cultivation of the two plants almost entirely, and later on the demand for real vanilla may be so much revived that the cultivation of this orchid again becomes profitable.

Certain drugs and condiments have the advantage of being eminently suited for an export trade because they are both light and expensive, like black pepper or saffron. This group of plants merits the fullest attention of the plant-breeders.

Triploidy

So far, we have dealt chiefly with examples in which polyploidy has given advantageous results, but we must now mention some of the difficulties that polyploids offer to the breeder—difficulties which may, however, be guarded against by reference to a list of chromosome numbers. When polyploids with the same numbers of chromosomes are crossed, the result is a hybrid which, as a rule, is of moderate fertility; but when their chromosome numbers are different, difficulties are to be expected at reduction with consequent loss of fertility. Take the case of a cross between a diploid with a single set of chromosomes in its gametes (like Prunus cerasifera, the myrobalan plum) and a hexaploid with three sets (like the "Jefferson" variety of P. domestica), such as that made by Crane.1 Irregularities of reduction might be expected. But four chromosome sets are brought together in the hybrid just as in the cross between Tritkum durum (4.x) and T. polonicum (4.x).

Three of these are derived from one parent and one from the other, instead of two from each. Does this matter, since in any case the four sets are originally derived from four different species ? So far as behaviour of the chromosomes is concerned it certainly does not. The hybrid behaves just like any other tetraploid ami even gives seedlings when back-crossed with the hexaploitl plum patent. The same kind of result has been achieved in Papaper, Crepis and Digitalis hybrids.This is well enough where there is an even number of sets of chromosomes in the hybrid, but where the number is odd, reduction can never be regular; chromosomes of the odd set have nothing corresponding to pair with, and they pass at random to one pole or the other. The result is the production of germ-cells with all kinds of chromosome numbers, and we know from other instances with odd chromosomes what this means. Here the case is worse, for a triploid ornamental cherry like Prunus avium var. nana (a cross between the diploid sweet cherry and the tctraploid sour cherry) has eight extra chromosomes. It produces germ-cells with every number between eight and sixteen, and among these the balanced ones with eight and sixteen only come once in over a hundred times. The result is that such a triploid is highly infertile. This, however, has its uses. The triploid watercress is sterile. It is eaten in winter and possibly is sturdier on account of its seed-sterility. The cherry in question is an ornamental variety, and if the cherries of the Japanese group are examined, it is found that nearly all of them are either triploid or double-flowered. Both these types are sterile. Clearly, gardeners have chosen for ornamental purposes those varieties which produce least fruit, for in trees which are grown for a display of flowers, fruit is often a disadvantage. Therefore, if one is raising ornamental trees which can be propagated readily by cuttings or grafts, tetraploids should be crossed with diploids to get, not merely something new, but also something with the advantage of sterility. Actually the triploid Japanese cherries have arisen accidentally, and not by design. Their sterility depends simply on the presence of an odd set which has been Introduced without hybridisation; the odd set is the result of a germ-cell with the unreduced diploid number on one side fusing with a haploid germ-cell on the other.

This is what has happened in the garden tulips. Several important varieties such as Keizerskroon, Massenet, Pottebakker and Pink Beauty, are triploid. They arose from the fusion of a diploid egg with a haploid pollen grain. Consequently, unlike the products of hybridisation with non-giant tetraploid species, such triploids usually show a suggestion of gigantism which has led to their selection.

This is best exemplified by the hyacinths, but it must be pointed out that the hyacinths are altogether exceptional in not being badly affected by having odd chromosome numbers. Therefore their triploids are not sterile, but give rise to seedlings having various numbers, which, although less vigourous than the triploids themselves, are occasionally worth growing. The best varieties of garden hyacinths are triploids. The first to appear was Grand Maitre in 1870.

They are all larger in bulb and flower than either the diploids which preceded them or the odd-numbered forms to which they have given rise as seedlings. They proliferate freely, their bulbils maturing in four years, that is more rapidly than those of the diploids.

SECONDARY POLYPLOIDY

The apple and pear group offer another interesting peculiarity in their chromosome properties, and therefore in their inheritance. It will be seen that the basic number 17 is not paralleled in any of the other examples that have been given. The question arises, Is this really a basic number, or does it conceal the evidence of a more complicated origin from a lower basic number ?

Close study of germ-cell formation in the apples and pears and medlars has shown that 17 is not the primary number, but is really derived from an earlier set of seven, like that found in roses and strawberries. This earlier set at a remote period was doubled in part and triplicated in the other part. If the earlier set of 7 was ABCDEFG, then the new set of 17 is AAA9 MB, CCC, DDy EE, FF9 GG. It is unbalanced relative to the old set of seven, and it may be that the apple group owes its special characteristics, more particularly its fruits, to this change in balance. Another species which probably arose as an unbalanced polyploid and owes its special qualities to its new balance is Dahlia Merckii. This species differs from other dahlias in having thirty-six chromosomes instead of thirty-two or sixty-four. It is not in the eight-series, but has two extra pairs of chromosomes. It differs in form from all the other Dahlia species with multiples of eight by having fertile ray florets. It is another case of unbalance being a good balance.

Crossing Polyploid Species

If this is the way polyploid species behave when they ate continually inbred, something rather complicated is to be expected when they are crossed. Diploid species-crosses very often give sterile hybrids. Polyploid species, if they have the same number of chromosomes, cross more often to give fertile hybrids. Polyploid species are therefore often more difficult to separate than diploid. Since it is not known how the polyploid species themselves are made up (except that they are hybrid), it cannot be predicted how they will behave when they are bred together; the same combination may be broken up in many different ways.

One of the exceptional results they give is what Engle-dow has called "shift." In the offspring of a cross between Triticum polonicum and T. durum he found no segregation of the long-glume character of the Polish wheat. All the progeny for several generations remained intermediate in character. In the same way, in a cross between a rough-chaffed rivet wheat, T. turgidum, and T. polonicum, Biffen never recovered the grey chaff colour on the rivet, although progeny were grown to the number of 100,000. The same suppression of a character by crossing has been found by Backhouse and Vavilov in the tetraploid cereals, and in other cases they have found the reverse effect of a character being accentuated.Take the case of the cultivated strawberry, whose history have been well described by Bunyard and Pearl. The wood strawberry, Fragaria vesca, was cultivated for its fruits in the fifteenth century. It produced no variations of importance;, except an ever-flowering "alpine strawberry" analogous to the "All Saints" cherry and the "Lloyd George" raspberry. The fruits always remained small.

This species is diploid. Another species that came into cultivation at this time in France and England was Fragaria elatior, the Hautbois strawberry, also a native of Europe. This species differs from the former in holding its ripe fruits above the leaves, and in the musky flavour of the fruits. Like the wood strawberry, it has small fruits, and although still occasionally grown, it likewise has shown no considerable improvement in size in the course of four hundred years of cultivation. This species is usually dioecious, having male and female flowers on different plants. F. elatior is hexaploid, and therefore will not yield fertile hybrids with F. vesca.

In the seventeenth century, the scarlet strawberry, F.virginiana, from North America was introduced by John Tradescant. The plant was distinct in vegetative characters and its fruit was different in shape and flavour from that of the two European species, but it was scarcely any bigger, about half-an-inch long. It produced more varieties than the older strawberries, but no marked improvement. This species is octoploid and does not cross with either of the first two.

The last important species to be introduced to Europe was the South American F. chiloensis which came to France in 1712. But the plants introduced were all females, and consequently set no seeds until planted near other species with pollen. F. chiloensis—very distinct from the others, having stout scapes with few flowers, and fruits twice their size—is octoploid. From all accounts it is certain that the few plants of it that were introduced afforded the first possibility of raising the large-sized fruits familiar in our modern cultivated varieties. But since the plants were female, it is evident that the first varieties to occur in cultivation were the results of natural crossing with other species. All the modern large-fruited varieties examined are octoploid, and it is quite clear that they arose from crossing with the other octoploid species, F. virginiana.

Great numbers of improvements, like this in strawberries, have been found to occur following the introduction of species into cultivation, and. they have always been attributed to cultivation. If "cultivation" means "opportunity for hybridisation," the popular idea is correct. No other effect of cultivation has ever been shown to have any importance in plant improvement. It is curious that in spite of improved varieties of strawberries being raised from crossing, they were not sufficiently improved to suggest the advantage of further crossing. The only crosses, as we have seen, were accidental. In fact, we may say they could hardly have been helped. It was only a hundred years later, with the fashion set by Knight in England of raising plants from seeds, that new strawberry seedlings appeared. Soon after, by systematic crossing of the derivatives of these early seedlings, Laxton produced the first strawberries having the qualities required in cultivation to-day. The same kind of change is undoubtedly responsible for the improvement of the Chilean wild potato. But here part of the improvement was probably carried out by the natives in South America before introduction into Europe, and only part after.

In this way it is clear that almost as rapid improvement can be made by crossing polyploid species with the same chromosome numbers as by crossing diploids and raising polyploids from them. The result, too, will not depend on chance, for polyploidy is there to start with. When two tetraploid species are crossed, four different chromosome complements originally derived from four different species are thrown into the melting pot. Not only is the result often fertile, which might well not be, were the parents diploid, but an enormously increased number of combinations can be made from the sorting out of the chromosomes derived from the four ancestral species. Among these combinations the breeder can select the new and improved forms that he wants, and propagate them vegetatively as in the case of strawberries and potatoes; or, by inbreeding, he can extract new strains that will breed true, as in the cereals.

Conclusion

We have attempted to find, for the guidance of the plant-breeder, the materials that transmit the hereditary qualities of the plant from one generation to the next; we have concluded that these materials lie in the chromosomes. Study of the behaviour of the chromosomes has shown how their constancy is preserved and how they vary. Particularly we see that in their reduction the chromosomes provide the mechanism of Mendelian segregation and at fertilisation the mechanism of Mendelian recombination.

But in the various irregularities of their behaviour are to be seen the means by which types of inheritance can be developed much more complicated than any expressed in the simple Mendelian scheme, although they actually arise out of the same mechanism. Among these are the examples of true-breeding hybrids, such as the ring-forming evening primrose and the hexaploid bread wheat.

We also know now how sports arise from the presence of an extra chromosome and how the ever-sporting habit may be preserved by the loss of a piece of chromosome. We can see how many important improvements in plants were brought about by polyploids. We know how such plant breeders as Thomas Laxton and Thomas Andrew Knight got the results they did, although working relatively in the dark.

And we can thereforeimitate them in other fields which appear, in the light' of a knowledge of the chromosome numbers, as promising as those they worked in, The occasional seedlings of sterile diploids can be raised in order to obtain fertile polyploids. Where a sterile hybrid is no disadvantage diploids can be crossed with polyploids. We can see, moreover, that chromosome studies are particularly instructive in relation to fertility. Generational sterility is merely a symptom of segregation.

It is due to the pairing and separating of chromosomes that are not sufficiently like one another (as in the diploid hybrid Primula kewnsis), or to a less extent to chromosomes that are like one another pairing and failing to separate (as in tetraploid tomatoes). Further, there are many plants that are sterile, not on account of any hybridity in the ordinary sense, but simply because one of their parental germ-cells had two sets of chromosomes instead of one, so that they have an odd set which cannot pair normally at reduction.

The degree of sterility to be expected in a plant can therefore be calculated to some extent from the pairing of its chromosomes and the pairing can be predicted from a know-, ledge of the parents' chromosomes. The plant-breeder therefore is in a position to decide before making a cross whether the results are likely to be fertile or sterile, and this is of great importance in the work of raising new econo-mic plants. Finally, we can understand a great deal more of the nature of species, a knowledge of which is as important to the plant-breeder, from one point of view, as it is to the systematise from another. A study of the chromosomes shows at once that there are some species which are homosygous, uniform and true-breeding. These are the simple-pairing diploids. There are others, equally true-breeding and uniform, which are of hybrid origin and have certain properties, more or less clearly shown; of hybrids. These are the ring-forming diploids and hybrid polyploids. There are others again that contain several inter-sterile groups owing to polyploidy having arisen within them without hybridisation.

And finally there are some that are seed-sterile hybrids maintaining themselves as uniform types by vegetative means such as bulb formation, apomixis, and so on. Cytological examination at once shows which of these groups a species falls in, and is a necessary preliminary to any systematic breeding work. While chromosome studies therefore do not give the plant-breeder any greater control over his material, they enable him to direct his efforts into the right channels for obtaining and preserving the results likely to prove most profitable to him.

9

In vitro Techniques for Collection and Exchange of Germplasm

IN VITRO COLLECTING

Collectors are faced with various problems when collecting germplasm of recalcitrant seed and vegetatively propagated plant species. Collecting missions often require travelling for relatively long periods in remote areas. It is thus necessary to keep the material collected in good state for some days/weeks before it can be placed in optimal growth or storage conditions. There are thus great risks that recalcitrant seeds either germinate or deteriorate before they are brought back to the genebank.

In addition, many recalcitrant seeds have considerable weight and bulk, which increases the volume of material to handle and induces additional costs, if an adequate sample of the population is to be collected. With vegetatively propagated species, the material collected will consist of stakes, pieces of budwood, tubers, corms or suckers. Not only will most of these explants not be adapted to survival once excised from the parent plant, but they will also present health risks due to their vegetative nature and contamination with soil-borne pathogens.

Difficulties can also be encountered when collecting germplasm of orthodox seed-producing species. Even with careful planning of the time of a collecting mission, there might be no or little seed available for all or part of the germplasm to be collected, or seeds might not be at the optimal developmental stage, or might be shed from the plant or eaten by grazing animals.

The problems described previously can be overcome if it is realized that the seed is not the only material which can be collected: zygotic embryos, or vegetative tissues such as pieces of budwood, shoots or apices can be sampled, transported and grown successfully if placed under adequate conditions. Following an expert meeting organized by IBPGR in 1984 and sponsorship of various research programmes, *in vitro* collecting techniques have been developed for different materials including embryos of coconut, cacao, avocado,

Citrus, vegetative tissues of cacao, *Musa*, coffee, *Prunus*, grape, cotton and several forage grasses.The critical points to consider for the development of *in vitro* collecting techniques have been synthesized and analysed by Withers (1995). The techniques developed are very simple and flexible, as illustrated below with coconut and cacao.

In the case of coconut, the *in vitro* field collecting technique developed by Assy-Bah *et al.* consisted of extracting from the nut with a corkborer a plug of endosperm containing the embryo. After surface sterilization with calcium hypochlorite or commercial bleach, the embryos were dissected on the spot under the shelter of a wooden box and inoculated onto semisolid medium, or the endosperm plugs were transported to the laboratory where the embryos were dissected and inoculated onto semisolid medium in aseptic conditions.

This technique is very efficient since after approximately 6 to 9 months in culture in the laboratory under standard conditions, an average of 75% of the embryos collected developed into plantlets which could be successfully transferred to the nursery and then to the field. In addition, embryos inoculated *in vitro* in the field could be kept in the open for two months before being grown in the laboratory, without influencing their further development.

This technique has been modified by other researchers towards higher or lower sophistication. In its more sophisticated version, inoculation of the embryos onto sterile medium is performed in an inflatable glove box. The simpler procedure consists of sterilizing the endosperm plugs containing the embryos and of placing them in a cool box in sterile plastic bags for transportation. Upon arrival in the laboratory, a second disinfection is performed and the embryos are dissected and inoculated onto culture medium inside the laminar airflow cabinet.

In the case of cacao, an *in vitro* collecting method was developed for budwood. Considering that absolute sterility would be difficult to achieve in the field and would not necessarily be essential for robust, woody material, the aim was to place the samples collected in conditions which would suppress or delay deterioration. Stem nodal cuttings were disinfected with boiled water in which drinking-water sterilizing tablets had been dissolved and containing fungicides, then inoculated onto semisolid medium supplemented with fungicide and antibiotics. Explants could be maintained in a relatively clean (though not necessarily completely sterile) condition for up to 6 weeks.

In Vitro Germplasm Exchange

In vitro culture techniques have been used extensively for the international exchange of germplasm because of their obvious advantages over *in vivo* material, notably their reduced weight and volume, and phytosanitary condition. Indeed, meristem culture, alone or employed in combination with thermotherapy, can eliminate viral pathogens. Plant material can thus be multiplied, stored and exchanged in a disease-free state.

In particular, because aseptic culture conditions are used and provided that the plant material has been cleansed of internal contaminants, problems with the international movement of germplasm are considerably reduced. Most countries will accept batches of plants *in vitro* with a phytosanitary certificate, without requiring a rigorous quarantine period.

However, an important caution in this area is that transfer to *in vitro* culture does not confer disease-free status. Indexing, using one or a combination of the various techniques available, which include symptomatology, grafting/ inoculation on indicator plants, ELISA and molecular techniques such as dsRNA detection, is the only sure way of making this judgement.

Routine procedures for the international exchange of *in vitro* cultures have been developed, notably for potato, cassava, yam and *Musa*. Samples are usually placed in glass, or preferably plastic, test tubes on the standard culture medium, possibly with the gelling agent modified to increase its firmness and a reduction in the carbon source to limit growth.

Heat-sealable polythene bags are sometimes employed instead of more fragile plastic or glass test tubes. With species such as potato and sweet potato, where it is possible to induce their formation *in vitro*, tubers are also used for germplasm exchange, since they are often easier to transport and to handle than plantlets.

Encapsulation of plant material in alginate beads has been suggested recently as a possible way of germplasm exchange in the case of banana. Using encapsulated apices instead of *in vitro* plantlets would result in a further volume reduction.

In addition, encapsulated apices might be either grown *in vitro* after their receipt, or directly sown in the soil as seeds where they would develop into plantlets.

The encapsulation technique has also been employed with nodal segments of yam. In a germplasm exchange simulation experiment, encapsulated nodal segments were conserved for two weeks in the dark in cryotubes containing culture medium, without any detrimental effect on their further growth.

The most widely used method of conserving plant genetic resources depends on seed. Many species produce seeds that can be dried to low moisture contents and stored at low temperature. Their longevity can be extended by reducing their moisture content and decreasing their storage temperature. Seeds that can tolerate extensive desiccation and can be stored in this way are termed orthodox.

However, several categories of crops present problems with regard to seed storage. A number of species, predominantly tropical or subtropical, such as coconut (*Cocos nucifera*), cacao (*Theobroma cacao*) and many tree and shrub species, have seeds which do not undergo maturation drying and are shed at relatively high moisture contents. These seeds are unable to withstand much desiccation and are often sensitive to chilling, and therefore, cannot be stored dry at low temperature. These so-called recalcitrant seeds have to be kept in moist, relatively warm conditions, and even when stored in an optimal manner, their longevity is limited to weeks, occasionally months. Recent investigations have identified species which exhibit an intermediate form of seed storage behaviour. These seeds can tolerate desiccation to fairly low moisture content but the dry seeds are injured by low temperature. In comparison to truly recalcitrant seeds, the storage life of these seeds can be prolonged by some drying, but long-term conservation, i.e. comparable to orthodox seeds, remains unattainable. This category of seeds includes economically important species such as coffee (*Coffea* spp.) and oil palm (*Elaeis guineensis*).

Some crop species have genotypes which do not produce seeds and others, such as potato (*Solanum tuberosum*), yam (*Dioscorea* spp.), cassava (*Manihot* spp.), sweet potato (*Ipomea batatas*) and sugar-cane (*Saccharum* spp.), have either sterile genotypes or produce orthodox seeds which are highly heterozygous and are therefore of limited interest for the conservation of particular gene combinations. These species are mainly propagated vegetatively to maintain clonal genotypes. Other crop species such as banana and plantain (*Musa* spp.) do not produce seeds and are thus multiplied vegetatively.

At present, the most common method to preserve the genetic resources of these problem crop species is as whole plants in the field. There are, however, several serious problems with field genebanks. The collections are exposed to natural disasters and attacks by pests and pathogens; moreover, labour costs and the requirement for technical personnel are very high. In addition, distribution and exchange from field genebanks is difficult because of the vegetative nature of the material and the greater risks of disease transfer. Until now, most plant genetic resources conservation has focused on crop species. However, the conservation of rare and endangered plant species has also become an issue of concern.

Finally, the development of biotechnology has led to the production of a new category of germplasm, including clones obtained from elite genotypes, cell lines with special attributes, and genetically transformed material. This new germplasm is often of high added value and very difficult to produce. The development of efficient techniques to ensure its safe conservation is therefore of paramount importance. During the last 20 years, *in vitro* culture techniques have been extensively developed and applied to more than 1000 different species.

Tissue culture techniques are of great interest for the collecting, multiplication and storage of plant germplasm. Tissue culture systems allow propagation of plant material with high multiplication rates in an aseptic environment. Virus-free plants can be obtained through meristem culture in combination with thermotherapy, thus ensuring the production of disease-free stocks and simplifying quarantine procedures for the international exchange of germplasm. The miniaturization of explants allows reduction in space requirements and consequently labour costs for the maintenance of germplasm collections.However, the high multiplication rates which can be achieved using *in vitro* culture procedures lead to the regular production of large amounts of plant material. This creates problems for the management of large *in vitro* collections. In addition, risks of losing material through contamination or human error are present at each subculture. More importantly, the risks of losing the genetic integrity of the plant material through somaclonal variation increase with time in culture. Storage techniques which reduce the burden placed upon all *in vitro*-based procedures and preserve the genetic integrity of the plant material are urgently needed.

Different *in vitro* conservation methods are employed, depending on the storage duration required. For short- and mediumterm storage, the aim is to reduce growth and to increase the intervals between subcultures. For long-term storage, cryopreservation – *i.e.* storage at ultra-low temperature, usually that of liquid nitrogen (2196°C) – is the only current method. At this temperature, all cellular divisions and metabolic processes are stopped. The plant material can thus be stored without alteration or modification for a theoretically unlimited period of time. Moreover, cultures are stored in a small volume, protected from contamination, requiring a very limited maintenance. *In vitro* collecting, slow growth and cryopreservation techniques are described and analysed in the following sections. The sections describing *in vitro* and slow growth techniques have been adapted from a recent review by Withers and Engelmann (1997).

CRYOPRESERVATION

The development of cryopreservation for plant cells and organs has followed the advances made with mammalian species, albeit several decades later. The first report on survival of plant tissues exposed to ultra-low temperatures was made by Sakai in 1960 when he demonstrated that very hardy mulberry twigs could withstand freezing in liquid nitrogen after dehydration mediated by extra-organ freezing. More than a decade later it was shown that cultured cells of flax could resist freezing to 250°C after pretreatment with dimethylsulphoxide (DMSO).

This study was followed by a similar one in which cell cultures of carrot were shown to survive after freezing in liquid nitrogen. The methodology employed in the above experiments followed the classical procedures which had been successful with other living systems: chemical cryoprotection, slow dehydrative cooling followed by rapid immersion in liquid nitrogen, storage in liquid nitrogen, rapid thawing, washing and recovery. The development for plant tissue cultures of so-called classical techniques which took place in the 1970s and 1980s is based on the above sequence of treatments. In recent years, several new cryopreservation techniques have been developed which allow cryopreservation to be applied to a larger range of tissues and organs, in different technical environments.

Dehydration and Freezing Injury

Most of the experimental systems employed in cryopreservation (cell suspensions, calluses, shoot tips, embryos) contain high amounts of cellular water and are thus extremely sensitive to freezing injury since most of them are not inherently freezing-tolerant. Cells have thus to be dehydrated artificially to protect them from the damage caused by the crystallization of intracellular water into ice. The techniques employed and the physical mechanisms upon which they are based are different in classical and new cryopreservation techniques. Classical techniques involve freezeinduced dehydration, whereas new techniques are based on vitrification. Vitrification can be defined as the transition of water directly from the liquid phase into an amorphous phase or glass, whilst avoiding the formation of crystalline ice.

Classical cryopreservation techniques involve slow cooling down to a defined prefreezing temperature, followed by rapid immersion in liquid nitrogen. With temperature reduction during slow cooling, the cells and the external medium initially undergo supercooling; this is followed by ice formation in the medium.

The cell membrane acts as a physical barrier and prevents the ice from seeding the cell interior, so the cells remain unfrozen but supercooled. As the temperature is further decreased, an increasing amount of the extracellular solution is converted into ice, thus resulting in the concentration of extracellular solutes. Since cells remain supercooled and their aqueous vapour pressure exceeds that of the frozen external compartment, cells equilibrate by loss of water to external ice.

In optimal conditions, most or all intracellular freezable water is removed, thus reducing or avoiding detrimental intracellular ice formation upon subsequent immersion of the specimen in liquid nitrogen. Rewarming should be as rapid as possible to avoid the phenomenon of recrystallization, in which ice melts and re-forms at a thermodynamically favourable, larger and more damaging crystal size.

New techniques are vitrification-based procedures. In such procedures, the freezable intracellular water is removed prior to freezing by exposing the samples to highly concentrated cryoprotective media and/or air desiccation. Dehydration is followed in most cases by rapid cooling. As a result, the internal solutes vitrify and deleterious intracellular ice formation is avoided. Glass transitions (*i.e.* changes in the structural conformation of the glass) during cooling and rewarming have been recorded with various materials using thermal analysis.

Classical Procedures

For each new material to be cryopreserved, optimal conditions have to be defined for each of the successive steps of the protocol.

The physiological state of the material can affect its survival. Cell suspensions are more likely to withstand freezing when they are employed during their exponential growth phase, *i.e.* when they are small and have a relatively low water content. Survival of carnation shoot tips after freezing decreased progressively with their rank on the shoot apex, starting from the terminal meristem, thus reflecting physiological and developmental differences found in explants sampled on the same plant. Harding *et al.* (1991) indicated that a long-term period in tissue culture before cryopreservation significantly reduced the ability of potato apices to survive freezing. A pregrowth period before the cryoprotective treatment on medium with osmotically active compounds such as mannitol or sorbitol can increase freeze-tolerance.

Samples are submitted to a cryoprotective treatment before freezing. It is carried out using various cryoprotective substances such as dimethylsulphoxide, sorbitol, mannitol, sucrose or polyethylene glycol. Additional information on the role and mechanisms of action of cryoprotectants can be found in several reviews. Cryoprotectants are employed alone or in binary or ternary mixtures. Mixtures of cryoprotectants have proved especially effective with cell suspensions.

Classical cryopreservation procedures include a two-step freezing: slow, controlled cooling to a defined prefreezing temperature followed by rapid immersion of samples in liquid nitrogen. Freezing rates of between 0.5 and 2°C min21 down to prefreezing temperatures around 240°C generally give satisfactory results in most cases. However, while for some materials, such as oil palm somatic embryos, a wide range of freezing rates can be employed without modification of the recovery rate, other materials (e.g. grape cell suspensions) require very precise freezing parameters to achieve survival. Most classical freezing procedures require the use of expensive programmable freezing devices which achieve precise freezing conditions. However, sophisticated apparatus is not always necessary to obtain high survival. Withers and King (1980) have successfully cryopreserved various cell suspensions with an improvised and simple apparatus that can offer reproducible but Non-linear slow cooling. More recently, several authors have successfully employed domestic or laboratory deep-freezers to perform the slow cooling step.

Removal of cryoprotectants by washing after thawing, which was widely adopted in the past, was found to be deleterious for a number of cell cultures. A much less stressing technique for removing cryoprotectants, developed by Chen *et al.* (1984), consists of moving cultures on a filter paper through a series of Petri dishes of solid culture medium. This technique was successfully applied to various cell suspensions and calluses. In some cases, standard culture conditions have to be modified to improve recovery of cryopreserved cultures. Recovery of cell suspensions is generally improved if they are cultured for some days on solid medium before being transferred to liquid medium. Cultures are often placed in the dark or under reduced light intensity to avoid photooxidation phenomena which are harmful to the material. Finally, the culture medium can be transitorily altered by modifying its hormonal balance, mineral composition or by incorporating activated charcoal.

Classical freezing techniques have been assessed with different materials including cell suspensions, calluses, apices and embryos. They have proven to be highly successful with most cell suspensions and calluses, *i.e.* culture systems which consist of small units of relatively uniform morphology. However, apart from exceptions like carnation apices, these techniques are not suitable for the cryopreservation of larger units comprising a mixture of cell sizes and types, such as apices and zygotic and somatic embryos.

New Techniques

New cryopreservation techniques offer practical advantages in comparison to classical ones. Vitrification-based procedures are operationally less complex than classical ones since they do not require the use of a programmable freezer. In addition, since ice formation is avoided during freezing, they are more adapted for freezing complex organs such as apices or embryos which contain a variety of cell types, each with unique requirements under conditions of freeze-induced dehydration. Finally, they have greater potential for broader applicability than classical techniques since only minor modifications are required for various cell types. Seven different vitrification-based procedures can be identified: (i) encapsulation-dehydration; (ii) a procedure actually termed vitrification; (iii) encapsulation-vitrification; (iv) desiccation; (v) pregrowth; (vi) pregrowth-desiccation; and (vii) droplet freezing. Table presents a list of species to which these new freezing techniques have been applied.

Encapsulation-dehydration

The encapsulation-dehydration technique is based on the technology developed for the production of synthetic seeds where somatic embryos are encapsulated in a bead of hydrosoluble gel. This technique has been applied mostly to apices of more than ten species of temperate and tropical origin, and to somatic embryos of several crop species as well as to a *Catharanthus* cell suspension.

Before cryopreservation, specimens can be submitted to conditioning treatments which increase their survival potential. In the case of cold-tolerant species, *in vitro* mother plants or apices can be placed at low temperature for several weeks. Mulberry apices are transferred daily on media with progressively increasing sucrose concentrations to initiate dehydration.

Table. List of plant species for which new cryopreservation techniques (encapsulation-dehydration,Vitrification,encapsulation-vitrificatio,pregrowth=desiccation,pregrowth,desiocation)have been developed using different types of Specimens.

Species	Specimen	No.of Accessions	Technique	
Aesculu				
Hypocastanea	Zygotic embryo	1	Desiccation	
Allium Sativum	Apex	12	Vitrification	
Allium wakegi	Apex	7	Vitrification	
Arachis hypogaea	Zygotic embryo	6	Desiccation	
Araucaria excelsa	Zygotic embryo	1	Desiccation	
Armoracia rusticana	Hairy root culture	1	Encaps./.	
dehyde.				
		1		
Artocarpus				
Heterophyllus	Zygotic embtyo	?	Desiccation	
Asparagus				
officinalis	Stem segment	1	Pregrowth/ desicc.	
	somatic embryo	1	vitrification	
	Cell suspension	1	vitrification	
		1		
Baccaurea motleyana	Zygotic embryo	1	Desiccation	
Baccaurea polyneura	Zygotic embryo	1	Desiccation	
Beta Vulgaris	Apex	2	Encaps./dehydr.	
Brassica campestris	Cell suspension	1	Vitrification	
Brassica napus	Microspore desicc.	1	Pregrowth/	Embryo
		1	Encaps/dehyd.r	
Bromus inermis	Cell suspension	1	Vitrification	
Calamus manan	Zygotic embryo	1	Desiccation	

Specimens are usually encapsulated in 3% calcium alginate gel. They are then submitted to the following successive steps: pregrowth, dehydration, freezing, thawing and recovery. Pregrowth is performed in liquid medium enriched with sucrose (0.3 to 1.5 M) for periods varying between 16 hours and 10 days. Replacement of sucrose with other sugars did not improve survival of grape apices. Progressive increase in sucrose concentration generally overcomes sensitivity to direct exposure to high sucrose levels which is encountered with some species such as eucalyptus, grape and coffee.

Desiccation is performed using either the air current of a laminar airflow cabinet or silica gel. The latter method is generally preferred because it ensures more precise and reproducible desiccation rates. The water content of the beads allowing optimal survival rates is around 20% (fresh weight basis). If pregrowth conditions have been well defined, only limited loss is observed after desiccation with most species. However, banana apices were found to be highly sensitive to both sucrose pregrowth treatment and to even limited desiccation, which induced a drastic survival drop. Desiccated samples are usually frozen by direct immersion in liquid nitrogen. However, modifications in the freezing rate can have different consequences on the survival of different materials. Survival of encapsulated grape apices was higher after slow cooling down to 2100°C. In contrast, survival of sugar-cane apices was higher after rapid freezing than after slow, controlled cooling. Freezing encapsulated carnation apices at cooling rates of between 0.5 and 200°C min21 had no effect on survival.

Samples are usually stored in liquid nitrogen at 2196°C. Scottez (1993) and Tannoury (1993) showed that survival of pear and carnation apices, respectively, was not modified after 2 and 3 years of storage at 2196°C. Several authors have demonstrated that samples can be frozen and conserved in deep-freezers provided that the storage temperature is below that of ice recrystallization (250 to 270°C). Pear apices were conserved at 275°C for one year and apple, pear and mulberry apices were stored for 5 months at 2135°C.

Samples are usually placed directly under standard conditions for recovery. However, transitorily modified conditions can enhance recovery in some cases. Sugar-cane apices are placed in the dark for one week on a medium supplemented with growth hormones. Addition of the antioxidant ascorbic acid significantly improved the recovery of sugar-beet apices, which are extremely sensitive to oxidation. Extraction of apices from the beads was necessary to allow regrowth of pear and grape apices. Regrowth of material frozen using the encapsulation- dehydration technique is usually direct and rapid, without callus formation. This is due to the fact that, contrary to what is observed after classical freezing, where many cells are destroyed and callusing is frequently observed during recovery, encapsulation-dehydration preserves the structural integrity of most cells. Therefore, regrowth usually originates from the whole meristematic zone, as observed notably in the case of sugar-cane.

There are presently four crops (pear, apple, sugar-cane and potato) on which the encapsulation-dehydration technique has been successfully extended to several genotypes or varieties. In all cases, even though genotypic variations were noted, results were sufficiently high to envisage routine application of the cryopreservation protocols developed.

Vitrification

Vitrification involves treatment ('loading') of samples with cryoprotective substances, dehydration with highly concentrated vitrification solutions, rapid freezing and thawing, removal of cryoprotectants ('unloading') and recovery. Vitrification procedures have been developed for more than 20 different species using protoplasts, cell suspensions, apices and somatic embryos.

The physiological state of the explants can influence their survival potential. In some cases, the plant material is thus submitted to various treatments before the cryopreservation procedure itself. *In vitro* mother plants can be cultured at low temperature for several weeks. Explants can be placed on a medium with cryoprotectants and/or cultured at low temperature for a short period. Pepó

(1994) indicated that axillary shoot tips are much more sensitive to vitrification than apical shoots. Loading consists of placing the explants in liquid medium containing cryoprotective substances (sucrose, glycerol, ethylene glycol) for a short period, varying between 5 and 90 min, depending on the material. This reduces the sensitivity of the material to the highly concentrated vitrification solutions.

Vitrification solutions are complex mixtures of cryoprotectants which have been formulated for their high ability to vitrify (*i.e.* form an amorphous glassy structure). Most solutions employed are derived from ones elaborated by either of two groups: that of Sakai's group, which consists of 22% glycerol, 15% ethylene glycol, 15% polypropylene glycol, 7% DMSO and 0.5 M sorbitol; and that of Steponkus' group, which comprises 40% ethylene glycol, 15% sorbitol and 6% bovine serum albumin. The duration of contact between explants and the vitrification solutions is a critical parameter, in view of their high toxicity. The dehydration period generally increases with the size of the explants used. Rye protoplasts are dehydrated for 60 s only, whereas the optimal dehydration duration of apices of pear and apple is 80 min.

Performing the dehydration step at 0°C instead of room temperature reduces the toxicity of vitrification solutions and thus broadens the window of exposure durations ensuring survival of samples. This also allows the manipulation of a large number of samples at the same time. Survival of asparagus embryogenic cell suspensions dropped rapidly after 5 min of dehydration at 25°C but high survival was obtained for dehydration periods between 5 and 60 min if it was performed at 0°C.

Specimens are then cooled rapidly by direct immersion in liquid nitrogen in order to achieve vitrification of internal solutes. Reduction in the volume of cryoprotectants and the use of plastic straws (500 ml–1 ml) which have a small diameter and a large surface area of contact with the exterior allow an increase in the cooling rate, reaching 990°C min21 in the case of asparagus cell suspensions frozen in 50 ml of medium in a 500 ml straw. Apices of mint and sweet potato were frozen without cryoprotective medium, thus reaching a cooling rate of 4800°C min21.

Rewarming of samples is performed as rapidly as possible to avoid devitrification, which would lead to the formation of ice crystals detrimental to cellular integrity. Samples are immersed in a water-bath or liquid medium held at 20–40°C.

Unloading aims at removing progressively the vitrification solution in order to reduce the osmotic shock. Liquid medium containing 1.2 M sucrose or sorbitol is added progressively to dilute the vitrification solution. Explants are then transferred to standard conditions. Vitrification procedures generally achieve high survival rates. Recovery is usually direct and rapid, even though a few authors have reported callusing and/or abnormal plant development. Vitrification experiments involving a large range of genotypes have been performed with several species including *Allium*, tea, *Citrus*, apple, mulberry, grape, potato and wasabi.

Encapsulation-vitrification

This technique is a combination of encapsulation-dehydration and vitrification procedures. Samples are encapsulated in alginate beads, then subjected to freezing by vitrification. It has been developed by Tannoury *et al.* (1991) with carnation apices, and applied recently by Sakai's group to apices of *Armoracia*, lily and wasabi.

In the case of carnation, encapsulated apices were pregrown for 16 h with progressively more concentrated sucrose solutions, then incubated for 6 h in a vitrification solution containing ethylene glycol and sucrose, and frozen either rapidly or slowly. Maximum survival was 100% and 92% after rapid and slow cooling, respectively. The technique employed with wasabi apices was slightly different. Apices were encapsulated in beads containing various loading solutions (glycerol or ethylene glycol and sucrose), then placed in the vitrification solution either at 25°C for 30 min or at 0°C for 70–100 min before rapid freezing. Ninety-five per cent of frozen apices recovered growth within 3 days.

Even though this technique has been used with a limited number of species only, it possesses great potential both in terms of efficiency and practicality. Matsumoto *et al.* (1995) mention that the recovery rate of apices frozen using the encapsulation-vitrification technique was 30% higher than with the encapsulation-dehydration technique. A reason for this might be that the alginate capsule reduces the toxicity of the vitrification solution. From a practical point of view, the number of manipulations is reduced as well as the total duration of the procedure.

Desiccation

Desiccation is a very simple procedure since it consists of dehydrating the plant material before rapid freezing by direct immersion in liquid nitrogen. This technique has been applied mainly to zygotic embryos or embryonic axes of various species, including numerous tropical forest trees, as well as to somatic embryos of several species and to shoot tips of mulberry.

Various parameters, such as the developmental stage of the embryos at the time of harvest, can greatly influence survival. Mature coffee embryos displayed a higher survival rate than immature ones. High variability was observed in the survival of embryos of several recalcitrant-seed-producing tree species harvested at different periods.

Desiccation is usually performed by placing the embryos in the air current of a laminar airflow cabinet. However, more precise and reproducible desiccation conditions are achieved by placing the embryos in a stream of compressed air or in an airtight container with silica gel. Optimal survival rates are obtained when the water content of the embryos is around 10–20% (fresh weight basis). Freezing is usually rapid but positive results have been obtained using slow cooling with several tropical forest tree species.

Regrowth of the frozen material usually takes place in standard conditions even though modification of the hormonal balance of the recovery medium proved beneficial with coffee zygotic embryos. Direct development of the embryos into plantlets is common but abnormal development patterns have been noted in some cases, such as the Non-development of the haustorium in the case of *Howea* and *Veitchia*, or callusing and/or incomplete development with *Hevea*, *Castanea* and *Quercus*, and oil palm.

Pregrowth

This technique consists of cultivating the plant material for different durations (hours to weeks, depending on the material) in the presence of cryoprotectants, then freezing rapidly by direct immersion in liquid nitrogen. It was first developed by Monnier and Leddet (1978) with zygotic embryos of *Capsella bursa-pastoris*, then applied to embryos of bean, wheat and maize, and more recently to meristematic clumps of banana.

Culture for 3 days up to 3 weeks on solid medium with high sucrose concentration (0.3 to 0.9 M) was employed with *Capsella,* maize and banana whereas a short treatment with liquid medium containing polyethylene glycol, glucose and DMSO was sufficient to ensure survival of bean and wheat zygotic embryos.

The pregrowth technique was successfully applied to five banana cultivars, resulting in survival rates ranging between 6 and 42.5%. Modifications in the sucrose concentration and the duration of the pregrowth treatment should lead to improvements in the survival rate.

Pregrowth-desiccation

Pregrowth-desiccation is a combination of the two techniques described immediately above. In this technique, samples are treated with cryoprotectants, partially desiccated, cooled and rewarmed rapidly. This technique has been applied to a limited number of specimens only: somatic embryos of oil palm, date palm, coffee, pea and melon; stem segments of *in vitro* plantlets of asparagus; microspore embryos of rapeseed; and zygotic embryos of coconut.

Culture of samples on media containing cryoprotectants usually takes place before desiccation. However, coconut zygotic embryos are dehydrated before preculture with cryoprotectants (Assy-Bah and Engelmann, 1992a). Sugars (sucrose, glucose) are generally employed for preculture. The duration of the preculture varies between 20 h for coconut and 7 days for oil palm. Various methods are employed for desiccation: coconut embryos are placed in the air current of a laminar airflow cabinet; asparagus stem segments and oil-palm somatic embryos are dehydrated using silica gel; melon somatic embryos are placed over a saturated salt solution; and date palm, coffee and pea somatic embryos are dehydrated using a stream of compressed air. The optimal water content (fresh weight basis) varies between 11.8% for melon and 25–30% for oil-palm somatic embryos.

Desiccated samples are frozen rapidly by direct immersion in liquid nitrogen. Specimens are usually stored in liquid nitrogen. However, Dumet *et al.* (1994) have demonstrated that oil-palm somatic embryos could be conserved for 6 months in a deep freezer at 280°C, *i.e.* below the glass transition temperature, without any modification in their recovery rate.

After rewarming, samples are usually placed directly under standard culture conditions. However, oil-palm somatic embryos are cultured on media with progressively reduced sucrose concentration, with transitory addition of 2,4-dichlorophenoxyacetic acid (2,4-D) to stimulate reproliferation.

Recovery of samples is usually rapid and direct. Alterations in the regrowth pattern of cryopreserved specimens were observed with coconut and oilseed rape microspore embryos. The haustorium of cryopreserved coconut embryos browned rapidly and did not develop further. Only half of frozen rapeseed microspore embryos developed directly into plantlets, whereas the other half produced callus and/or secondary embryos.

Pregrowth-desiccation has been tested with four varieties of coconut, giving recovery rates ranging between 33 and 93%. This technique is routinely employed for the longterm storage of 80 clones of oil-palm somatic embryos.

Droplet Freezing

The droplet freezing technique has presently been applied to potato apices only. Apices are pretreated for 2–3 h in liquid medium supplemented with DMSO, placed on an aluminium foil in 5 ml droplets of cryoprotective medium and frozen rapidly by direct immersion in liquid nitrogen. This procedure is adapted from the classical procedure developed by Kartha *et al.* (1982) with cassava where apices placed in droplets of liquid medium were frozen slowly in a programmable freezer. The droplet freezing method has been successfully applied to 100 varieties of potato with recovery rates ranging between 5 and 100%. Around 25 000 apices are stored for the long term in liquid nitrogen at DSM, Braunschweig, Germany.

Genetic Stability of in Vitro Conserved Material

Genetic conservation is based on the assumption that the material is stored under conditions ensuring genetic stability. However, there are various factors linked with *in vitro* culture/conservation procedures which can be a source of variation. These factors, as well as the various approaches for assessing the genetic stability of plants recovered from *in vitro* culture, have been reviewed recently by Harding (1995). Genetic variation can pre-exist in the collected material, possibly linked to its genetic structure, such as in sugar-cane or banana where polyploids are more prone to instability than diploids. It can also arise from somaclonal variation, particularly if the propagation system includes a dedifferentiated phase or from stresses imposed by the conservation procedures. The information available on the genetic stability of plant material conserved *in vitro* is presented below.

Slow Growth

Different genotypes do not grow at the same rate when placed in culture. This induces a risk of selection when plants are placed under stress conditions, such as slow growth. It is thus important to minimize the risk of selection. One of the most efficient ways is to use, whenever possible, differentiated, organized structures such as shoot tips or embryos for which the risks of variation are minimal.

A limited number of experiments only have been performed to assess the genetic stability of plants during slow-growth storage. The most exhaustive study has been performed in the framework of a joint project between CIAT and IPGRI which aimed at monitoring the genetic stability of cassava shoot cultures stored *in vitro*. Isozyme and DNA analysis as well as examination of the morphology of plants regrown in the field did not reveal any modification after 10 years of storage under slow growth. In contrast, growth of potato shoot tips on mannitol-supplemented medium, which caused morphological changes in the material, was correlated with DNA hypermethylation, which may be an adaptive response to conditions of high osmotic stress. This clearly illustrates the importance of defining storage conditions which are likely to cause as little stress as possible to the plant material.

Cryopreservation

Cryopreservation involves a series of stresses which might lead to modifications in the recovered cultures and regenerated plants. It is therefore necessary to verify that the genetic stability of the cryopreserved material is not altered before using this technique routinely for the long-term storage of germplasm. There is now increasing evidence that, provided the cryopreservation technique applied ensures the greatest possible maintenance of the integrity of the frozen specimen, there will be no modification at the phenotypic, biochemical, chromosomal or molecular level attributable to cryopreservation; none has been reported to date.

It has been shown that cell suspensions of numerous species maintain their biosynthetic and morphogenic potential after cryopreservation. The only exception concerns lavender cell suspensions exposed to successive freeze-thaw cycles, for which the number of colonies recovered from cryopreserved cells increased with the number of freeze-thaw cycles. However, no modifications were noted in the biosynthetic capacities of cryopreserved cells, suggesting a change in population structure rather than genetic change. A similar observation was made by Bercetche *et al.* (1990), who noted that cryopreserved *Picea abies* embryogenic calluses recovered faster than Non-frozen controls; Non-embryogenic tissues were killed by freezing, thus leading to the production of a more homogeneous population, with a higher embryogenic potential. This suggests that cryopreservation could be used as a tool to 'rejuvenate' cultures when their proliferation capacities are decreasing.

Plants regenerated from cryopreserved apices of strawberry and cassava were phenotypically normal. No differences were noted in the vegetative and floral development of several hundreds of oil palms regenerated from control and cryopreserved somatic embryos. In contrast, Fukai *et al.* (1994b) showed that a high percentage of plants regenerated from apices of a periclinally chimeric chrysanthemum cultivar frozen using a classical protocol had an altered flower colour. These apices had been severely harmed during freezing, leading to callusing during recovery. Regenerated plants had thus an adventitious origin which explained the disturbance in the initial chimeric structure. The same group compared the effect of freezing carnation apices using a classical protocol and the encapsulationdehydration technique. Encapsulation-dehydration ensured 100% recovery after freezing, and regrowth was rapid and direct. In contrast, after slow, controlled freezing, the recovery rate was 50% only, and callusing was observed during regrowth. The two above examples underline the importance of selecting the most appropriate freezing technique for any given material, not only as regards recovery rates but also as regards recovery pattern and genetic stability.

Electrophoretic profiles of two enzymatic systems were comparable in plants regenerated from control and cryopreserved apices of sugar-cane and sweet orange somatic embryos. The ploidy level of plants regenerated from oilseed rape somatic embryos and sensitive dihaploids of potato was not modified by cryopreservation. Restriction fragment length polymorphism (RFLP) patterns of plants regenerated from sugar-cane embryogenic cell suspensions were identical to those of unfrozen controls. Several molecular types were uncovered in plants recovered from both control and cryopreserved sugar-cane apices, thus indicating that the variation was not due to freezing but was Pre-existing among the *in vitro* mother plants. Finally, DNA analysis showed that a gene integrated in the genome of a navel orange cell suspension was maintained after freezing and one-year storage in liquid nitrogen.

Comparative Studies

A limited amount of work has been published on the comparative effects of slow growth and cryogenic storage on the stability of plant material. Mannonen *et al.* (1990) showed that the production of secondary metabolites by cell cultures of *Panax ginseng* and *Catharanthus roseus* decreased drastically after 6 months of culture either in standard conditions (*i.e.* with weekly subcultures) or in slow growth under mineral oil, whereas the productivity of cell suspensions cryopreserved and stored in liquid nitrogen during the same period was identical to that of the original cultures.

Modifications in the RFLP pattern were observed in potato plants stored for 6 months under slow growth on a medium supplemented with mannitol, whereas no such modifications were noted in plants regenerated from cryopreserved apices.

PRESENT USE OF IN VITRO CONSERVATION TECHNIQUES

Classical *in vitro* conservation techniques have been developed for a wide range of species, including temperate woody plants, fruit trees, horticultural species, as well as numerous tropical species.

However, a recent FAO survey indicated that only 37 600 accessions are conserved *in vitro* worldwide. Slow-growth conservation is used routinely for the conservation of only a few species, including banana, potato and cassava, in regional and international germplasm conservation centres such as INIBAP, CIP, CIAT and IITA.

Alternative medium-term conservation techniques are still at the experimental stage. Low-oxygen storage might be interesting for storing cold-sensitive tropical species since it allows growth reduction without decreasing the temperature. However, complementary experiments involving additional species and longer storage periods are still required. Medium-term storage of desiccated somatic embryos will be mainly applicable for the management of large-scale production of elite genotypes. Encapsulated apices might be employed for medium-term genetic resources conservation. Research is still needed to refine the protocols and extend the storage periods. Cryopreservation procedures have been developed for around 100 different plant species cultured in various ways including cell suspensions, calluses, apices, and zygotic and somatic embryos.

Most of this work has been performed in the framework of academic studies and has involved only one or a few genotypes. However, there are recent reports of large-scale experimentation of classical freezing techniques involving cell suspensions from several hundred genotypes. In the case of organized structures such as apices or embryos, the development of new cryopreservation procedures has also allowed experiments involving a relatively wide range of genotypes/varieties. In the case of potato, apices from more than 100 different varieties have been successfully frozen and around 25,000 apices are presently stored in liquid nitrogen.

There is an increasing number of cases where cryopreservation techniques can be considered operational. However, their routine application is mostly restricted to the conservation of cell lines in research laboratories. The only example of routine application of cryopreservation to another type of material is oil palm, where 80 clones of somatic embryos are stored in liquid nitrogen and samples thawed upon request for plant production.

SLOW-GROWTH STORAGE

Classical Techniques

Standard culture conditions can be used for medium-term storage with species which have a naturally slow-growing habit only. A wide range of *Coffea* species are thus conserved on standard medium at 27°C without subculturing for durations varying from 6 to 12 months, depending on the species. However, for most species, growth reduction is achieved by modifying the environmental conditions and/or the culture medium. The most widely applied technique is temperature reduction, which can be combined with a decrease in light intensity or culture in the dark.

Temperatures in the range of 0–5°C are employed with cold-tolerant species. Strawberry (*Fragaria* 3 *ananassa*) plantlets have been stored at 4°C in the dark and kept viable for 6 years with the regular addition of a few drops of liquid medium. Apple (*Malus domestica*) and *Prunus* shoots survived 52 weeks at 2°C. Tropical species are often cold-sensitive and have to be stored at higher temperatures, which depend on the cold sensitivity of the species. Kiwi fruit shoots can be conserved at 8°C and taro (*Colocasia esculenta*) tolerates 3 years of storage at 9°C. *Musa in vitro* plantlets can be stored at 15°C without transfer for up to 15 months. Other tropical species such as oil palm and cassava are much more cold-sensitive: oil palm somatic embryos and plantlets do not withstand even a short exposure to temperatures lower than 18°C. Cassava shoot cultures have to be conserved at temperatures higher than 20°C.

Various modifications can be made to the culture medium in order to reduce growth. Embryogenic cultures of carrot could be conserved on a medium without sucrose for two years, and reproliferated if a sucrose solution was supplied. Kartha *et al.* (1981) could conserve coffee plantlets on a medium devoid of sugar and with only half of the mineral elements of the standard medium.

Replacement of sucrose by ribose allowed the conservation of banana plantlets for 24 months. The addition of osmotic growth inhibitors (*e.g.* mannitol) or hormonal growth inhibitors (*e.g.* abscisic acid) is also employed successfully to reduce growth.

The type of explant as well as its physiological state when entering storage can influence the duration of storage achieved. Roxas *et al.* (1995) indicate that, in the case of chrysanthemum, nodal segments showed higher survival rates than apical buds. The presence of a root system generally increases the storage capacities, as observed by Kartha *et al.* (1981) with coffee plantlets. Microtubers can be successfully employed as storage propagules, as demonstrated with potato. Preconditioning the explants by exposing them briefly to temperature and light conditions intermediate between standard and storage conditions was favourable for *Nephrolepsis* and *Cordyline* cultures. Higher survivals were obtained with shoot cultures of wild cherry, chestnut and oak if they were kept for 10 days under standard conditions after the last subculture before their transfer to the cold storage chamber.

The type of culture vessel, its volume as well as the type of closure of the culture vessel can greatly influence the survival of stored cultures. Roca *et al.* (1984) indicate that cassava shoot cultures could be stored for longer periods in a better condition by increasing the size of the storage containers.

White spruce embryogenic tissues withstood a one-year storage period in hermetically sealed serumcapped flasks. Replacing cotton plugs by polypropylene caps, thus reducing the evaporation of the culture medium, increased the survival rate of *Rauvolfia serpentina* during storage. The use of heat-sealable polypropylene bags instead of glass test tubes or plastic boxes was beneficial for the storage of several strawberry varieties.

At the end of a storage period, cultures are transferred onto fresh medium and usually placed for a short period in optimal conditions to stimulate regrowth before entering the next storage cycle.

Alternative Techniques

Alternative techniques include modification of the gaseous environment of cultures, and desiccation and/or encapsulation of explants. Growth reduction can be achieved by reducing the quantity of oxygen available to the cultures.

The simplest method consists of covering the explants with paraffin, mineral oil or liquid medium. This technique was first developed by Caplin (1959), who could store carrot calluses for 5 months under a layer of paraffin oil. Augereau *et al.* (1986) and Moriguchi *et al.* (1988) applied this technique to calluses of *Catharanthus* and grape, respectively. Florin (1989) showed that 50% of a collection of 313 callus lines survived after storage under mineral oil for 12 months. However, Mannonen *et al.* (1990) reported that conservation under mineral oil did not preserve the productivity of *Catharanthus* and *Panax ginseng* callus cultures for more than 6 months.

Similar experiments performed with shoot cultures of various species led to contradictory results. After 4 months of storage under mineral oil, regrowth of surviving coffee shoot cultures was very slow and pear microcuttings did not survive. In contrast, several ginger species could be conserved for up to two years under mineral oil with high viability.

Reduction of the quantity of oxygen can also be achieved by decreasing the atmospheric pressure of the culture chamber or by using a controlled atmosphere. Tobacco and chrysanthemum plantlets could be stored under low atmospheric pressure (with 1.3% oxygen) for 6 weeks. Oil palm polyembryogenic cultures were conserved for 4 months at room temperature in a controlled atmosphere with 1% oxygen. Dorion *et al.* (1994) showed that hypoxic regimes at standard or intermediate temperature could replace low-temperature storage for shoot cultures of peach. Rice calluses could be stored for 12 weeks in a CO2 or N2 saturated atmosphere.

Desiccation of cultures as a means of achieving medium-term conservation was first reported by Nitzsche (1980), who stored dehydrated carrot callus for one year at 15°C under 25% relative humidity. Increasing interest has been paid recently to this technique, with the development of so-called synthetic seeds for various plant species. Synthetic seed is a generic term for a somatic embryo delivery system used as a means of clonal propagation. The aim is to use such somatic embryos as true seeds: embryos, encapsulated or not in alginate gel, could be stored after partial dehydration and sown directly *in vivo*.

After progressive dehydration using saturated salt solutions, naked alfalfa somatic embryos could be conserved with 10–15% moisture content at room temperature for one year and showed only a 5% decrease in their conversion rate at the end of the storage period. Carrot somatic embryos were stored for 8 months at 4°C without viability loss. Shorter storage durations were achieved with encapsulated material.

Encapsulated carrot somatic embryos survived after 3 months in liquid medium at low temperature and encapsulated shoot tips of *Valeriana wallichii* could be conserved for over 6 months at 4–6°C. According to Redenbaugh *et al.* (1991), the rapid dehydration of the encapsulating matrix limits the respiration of the encapsulated material, which is the cause for the rapid survival loss generally observed.

10

Genetic Resources and Plant Breeding

USES OF MARKER-AIDED SELECTION

In an ideal world, selection should be aimed directly at those genes which control the trait to be improved, but this assumes that genotypes at these target genes can be recognized easily, unambiguously and at an appropriate time without impeding the breeding programme. There are several reasons for using marker genes to improve selection efficiency.

Difficulties in Identifying Trait Genotype

The genotypes at a single-trait locus may be difficult to recognize because the desired alleles may have poor penetrance, be recessive, or interact with other genes or the environment. Environmental variation itself may impede accurate genotyping by causing the phenotype of different genotypes to overlap; this is particularly a problem with polygenic traits.

Some phenotypes such as resistance to a particular pest, pathogen or abiotic factor, like drought or salinity, may only appear under particular conditions which are difficult to define or control. These difficulties may be more severe during the early stages of a breeding programme when plant numbers are too few to allow adequate replication or the breeder does not want to place his valuable yet scarce breeding material at hazard.

Earlier Selection

Another major reason why it may be valuable to use marker or indicator genes is to reduce the time from sowing to selection. Many traits of economic importance are only apparent in the mature plant and so may not occur for months or even years after sowing. This is particularly a problem with biennial species and long-lived crops such as trees. Indicator genes, on the other hand, may be detectable in seedlings within days of sowing, if not in the seeds themselves, so avoiding the waste of valuable resources involved in raising and harvesting plants, most of which may prove to be of no value to the breeding programme.

More Intense Selection

Selection at a juvenile stage, particularly among seedlings, may also allow much larger populations to be studied and hence more intense selection to be applied. Indeed, selection may be possible in tissue culture without raising plants or running trials.

Non-destructive Scoring

Many characters are scored before maturity and the act of scoring may prevent seed being collected. Imposition of selection for pests and diseases may result in plants which are less able to produce seed. The use of certain types of marker loci may only require the removal of small quantities of leaf or other material to permit full genotyping of the plant.

Linkage Drag

Conventionally plant breeders have used backcrossing with selection to introduce a useful allele, such as one conferring disease resistance, from one strain into another. The source of the allele is often an alien or exotic species, but may be a related cultivar. The F1 is backcrossed to the recurrent parent, and backcrossing is repeated for five to ten generations, the intention being to introduce the desired donor allele whilst returning to the recurrent parent's genotype at all other loci. At every backcross generation, the breeder selects for the phenotype of the allele to be introgressed whilst simultaneously selecting for the phenotype of the recurrent parent for all other traits.

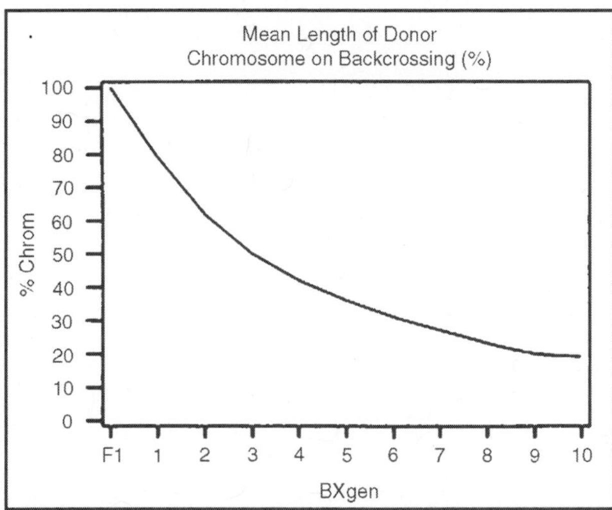

Fig. Linkage Drag.the Average Amount of Chromosome Surrounding a Selected GeneFollowing Backcrossing.Bxgen=Generation of Backcrossing from F_1

It was shown by Stam and Zeven (1981) that this procedure results in the leaving of very large amounts of donor genome associated with the selected alien allele. On average, about 20% of an average chromosome will consist of a donor segment even after ten backcross generations, while in any given line the actual size could be much more. Thus, as well as introducing useful alleles, the breeder will also be introducing very many undesirable alleles from the donor and so the gain in one trait may be at the expense of losses elsewhere. The use of marker loci can enable the breeder to reduce the size of this unwanted associated region as well as to accelerate the speed of return to the desired genotype of the recurrent parent

Heterosis Potential

Heterosis, or hybrid vigour, is due mainly to the presence of dispersed dominant alleles in the parental cultivars. Breeders concerned with developing lines for hybrid production are often concerned to identify lines which differ at as many gene loci as possible. It is often assumed that this implies that the lines also have very different genotypes at marker loci and hence such loci can be used to assess the genetic distance between potential parents, but this can be a dangerous assumption unless a wide range of different types of markers are used.

Quality Control

A major problem faced by most conservationists and breeders is to maintain the integrity of the material they are working with and to keep it free from contamination. Contamination can occur in various ways. It may arise through errors due to carelessness in labelling or sorting seed or plant material, mistakes in maintaining the pedigree, or as a result of outcrossing due to stray pollen from unrelated material. Such errors are difficult to recognize when only the gross phenotype is available as a guide, but genetic markers can provide a clear genetic fingerprint which can be used at any time to confirm the origin of the material. It can also be used to confirm that the intended cross or self has occurred, or to separate the desired progeny from a particular parent which might be a mixture of selfs and crosses.

Other Uses

Marker information can be used in other situations also: distinguishing haploids, diploids and aneuploids, for example, particularly following wide interspecific crosses or tissue culture. Doubled haploid lines produced by microspore or ovule culture can be recognized in this way and possible aneuploids arising during chromosome doubling can be identified and removed. Not only can genetic markers identify contaminant material but they also allow closely related or identical material to be identified in genebank collections, so reducing unnecessary duplication.

TYPES OF GENETIC MARKERS

Up to the mid-1960s the only easily usable genetic markers were those that produced clearly visible effects on the plant's phenotype such as colour, size, shape, disease resistance, etc.. All these were the result of mutations to alleles which had, obviously, a major effect on the development of the plant. Such mutants were rare and generally not desirable in commercial material although they may be selected by breeders of decorative plants for their novelty value. As genetic markers, however, they have many disadvantages. They often affect the fitness of the individual and could well have effects on important agronomic traits, *i.e.* they exhibit pleiotropy. Because they are rare, it would be necessary to bring several such mutants together in the same material in order to use them, and they would not normally be present in commercial cultivars or, indeed, in wild material. This not only takes time, but the combined effects of many such alleles would mitigate against their value as a genetic tool.

In the 1960s through to the 1980s, naturally occurring genetic polymorphisms at the protein level came to be easily recognized. Variation in enzymes and storage proteins was detectable on an electrophoretic gel for a very high proportion of gene loci which could be studied in this way and, unlike major mutants, such naturally occurring allelic variants had much smaller effects on fitness and hence were more common and existed naturally in populations. The number of such polymorphisms was, however, insufficient to provide the coverage of the genome desired by breeders and conservationists. The development of techniques during the late 1980s up to the present revealed the vast amounts of genetic polymorphism which exists at the DNA level and has revolutionized genetic analysis, opening up a whole new range of genetic tools.

As described elsewhere in this book, this variation arises through the existence of occasional base changes in the DNA which can be recognized by restriction enzymes or by primers used in analysis involving the polymerase chain reaction (PCR). Much of the DNA, possibly as much as 90% in many species, is Non-coding and hence reliance on morphological or biochemical variants was only using the restricted variation that plants could tolerate within the coding parts of their DNA.

Clearly, there are limits to the tolerance of such variation. Variation in the Non-coding regions, whether in the intergenic regions or within introns, is probably under far less constraint by natural selection, and hence the number of polymorphic sites is potentially enormous. Because most of these sites are outside coding genes they are referred to as loci rather than genes; genes are coding regions whilst most marker loci are not.

Depending on how the DNA polymorphism is studied, various types of marker are used. Restriction fragment length polymorphisms (RFLPs) have been the most widely used until now. They have the advantage of being codominant and hence all genotypes in a cross can be identified. On the other hand they require quite large quantities of DNA, and hence plant material, and generally rely on radioactive probes, although fluorescent techniques are becoming more popular.

As a result, RFLPs are expensive to use. The use of radioactive probes incurs safety considerations, as well as taking longer to visualize the bands. PCRbased markers are now being used more widely.

Randomly amplified polymorphic DNAs (RAPDs), for example, require far less material because only small amounts of DNA are needed and the required sequences are then amplified. They do not involve radioactive probes and are both quick and cheap. On the other hand RAPDs are dominant although it is possible to distinguish heterozygotes from homozygotes by the amount of product. Amplified fragment length polymorphisms (AFLPs) offer even cheaper and more easily identified and extensive PCRbased polymorphism.

Because of the number of polymorphisms which exist on individual gels and the speed of production, computer software is necessary to scan the banding patterns on the gels and to analyse and assimilate the information that is generated. Other popular markers are microsatellites (short sequence repeats) and cleaved amplified polymorphisms (CAP)-based primers.

All the genetic variants discussed above will be subsumed under the broad title of marker loci. They may be of little interest in their own right, and this is particularly true of the molecular polymorphisms; their value lies in their use as markers of more useful genes.

LOCATING GENES OF MAJOR EFFECT

The use of extensive molecular and other markers as described above provides a general framework map. Although this has value in its own right for comparing linkage groups in different species as well as possible structural variation within a species, the main value lies in providing a set of markers to locate genes of economic or special scientific interest.

Locating individual major genes is a straightforward development of the procedures above. Let us assume that a cultivar is identified which contains an allele of interest which changes the phenotype in a clearly recognizable fashion and which appears to segregate as a single gene in crosses to other cultivars. Using a knowledge of the positions of existing marker loci, a small subset of markers can be chosen which provide a reasonably even coverage of the genome, say every 20 to 30 cM, *i.e.* approximately five per chromosome. These then have to be shown to differ between the two cultivars and any which are monomorphic replaced by polymorphic markers.

A rough position of the gene of interest can then be obtained by bulked segregant analysis. This involves taking the mapping population, *e.g.* a Bc, and dividing the plants into two groups depending on whether or not they show the phenotype of interest. Bulk samples of DNA are taken from each group, and the two samples assayed for the marker systems chosen as in the paragraph above.

Any marker which is unlinked to the gene to be located will produce similar banding patterns in both samples because it will be segregating independently of the target gene. Conversely, should one of the markers be very closely linked to the target gene, it will co-segregate and the two samples will show quite different patterns for that marker.

In practice, complete co-segregation with any one marker is not very likely but, given a reasonable spread of markers over the genome, one or two are likely to show a strong association with the two groups. Thus although bands associated with both marker alleles will be found in each sample, their relative intensities will be quite different because of the association with the target gene.

Having located the chromosome and the region on the chromosome, the actual position can be identified in a segregating population using the principal marker associated in the bulked segregant analysis together with other polymorphic markers situated about 10 cM on either side of the principal marker.

The target gene is then mapped with respect to the three markers. This approach can be used for a wide variety of populations and traits. Even if the trait has poor penetrance, the affected group will clearly show the marker banding pattern even if the Non-affected group consists of a mixture due to misclassification.

MAPPING GENES CONTROLLING QUANTITATIVE TRAITS

Quantitative traits such as yield, quality, height and flowering time, which are controlled by several genes and are greatly influenced by the environment, create particular difficulties for gene mapping. The problems arise because the genotype for the trait concerned can never be clearly identified from the phenotype; many different genotypes could produce the same or very similar phenotypes whilst the same genotype can result in different phenotypes depending on what may be elusive factors in the environment.

Whereas with a single gene difference controlling a major effect, two or more Mendelian phenotypic classes might be recognized in ratios of 3:1, etc., quantitative traits resulting from the joint segregation of many genes show a continuous, often normally distributed range of phenotypes. The total phenotypic variation in an F2, for example, VP, is made up of genetic and environmental components, VG and VE. The genes underlying such traits are variously called polygenes, effective factors or, more recently, quantitative trait loci or QTL.

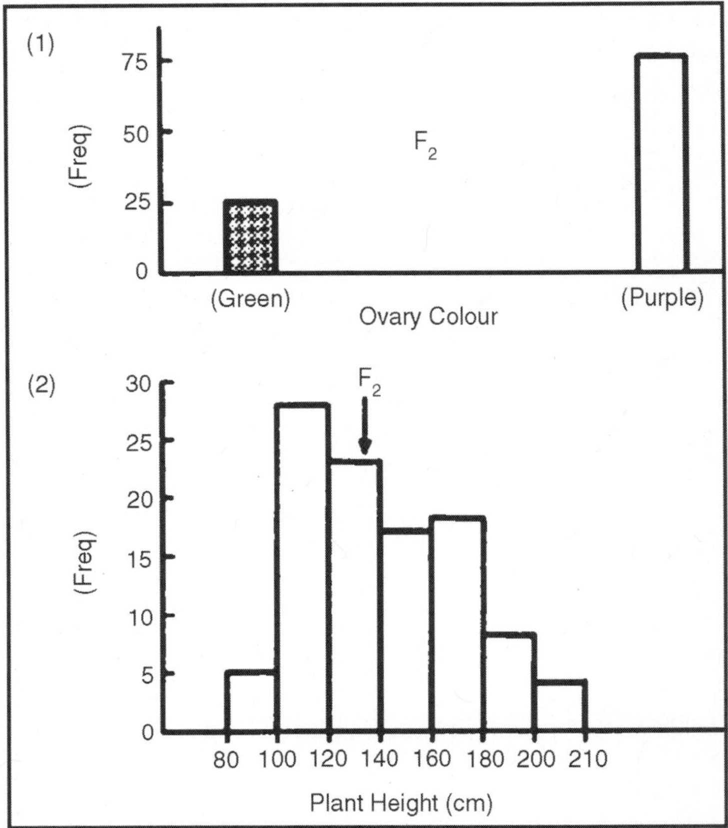

Fig. Contrast Between Distribution of a Qualitive and Quantitative trait in an F_2. Data for NIcotiana Rustica,Ovary Colour and Height.

However, it is still possible to map and measure the effects of the QTL by techniques analogous to those used for single major genes, *i.e.* by looking for the effects of co-segregation of particular marker loci with differences in the quantitative trait. Various mapping populations can be used as before, F_2, Bc, RIL, DH or open-pollinated full sib (FS) or half sib (HS) families, and the individuals or families are scored both for their phenotype for the quantitative trait(s) and for their genotype at the marker loci. The type of population that is used will depend on the traits studied and the breeding system of the plant species but, other things being equal, an F_2 population will normally provide the most information for a given size.

Unfortunately, most traits of economic importance such as yield, quality and disease resistance are not usefully scored in an F_2 population because one is interested in how the material performs in something similar to the high-density monoculture that it will encounter in agricultural practice. Measuring yield in an F_2 population with spaced plants is straightforward and statistically very powerful but it may well not provide any insight into how yield is controlled in agricultural practice. It is therefore necessary to use a plot structure for most traits in an attempt to simulate commercial conditions. For this reason most plant breeders will prefer to work with genotypes that can be extensively and easily replicated such as RILs, DHs or F_3s. It is also easier to explain the principles underlying QTL mapping using a population of DH lines, and so for both reasons the methodology will be illustrated with DH lines, although the same principles apply to all populations. Because several trait loci are concerned, probably on different chromosomes, bulk segregant analysis is of limited applicability, at least initially.

THEORY OF QTL MAPPING IN DOUBLED HAPLOID LINES

Doubled haploid (DH) lines can be produced parthenogenetically from an F1 by microspore or ovule culture and, in some species, such as wheat and barley, by wide outcrossing. Each original DH plant is derived from a single gamete, which subsequently becomes homozygous and disomic either naturally or by treatment with the drug colchicine, which inhibits spindle formation, and hence chromosome disjunction, at metaphase of mitosis.

Selfed seeds from such a plant produce a DH line of identical individuals so the line becomes effectively immortal. If we consider an F1 that is heterozygous for a marker locus M1/M2 and for a linked QTL locus Q1/Q2, where the 1/2 indicates the allele with an increasing or decreasing effect on the trait, this will produce the gametes and hence the DH lines shown in Table.

If the DH lines are scored for some trait for which the Q+/Q+ homozygotes increase the trait above the overall mean, m, by a units while the Q2/Q2 homozygotes decrease the mean by a, then the means of the four gametic types are as shown in Table, where R is the recombination frequency between the QTL and the marker. From this it follows that the mean trait score of all those DH lines which are M1/M1 (M1M1) is m 1 a(1 2 2R), while the mean of all those DH lines which are M2/M2 (M2M2) is m 2 a(1 2 2R). In other words, half the difference between these two means (M1M1 2 M2M2) is a(1 2 2R). Clearly, if the marker is so close to the QTL that it never recombines, R will be 0 and the marker difference will represent the QTL effect, a. Conversely if the two loci are unlinked, *i.e.* $R = 0.5$, then the difference will be zero. This effect is, of course, the basis of bulked segregant analysis. In general, the magnitude of the marker effect will decline linearly with the distance of the marker from the QTL in terms of recombination frequency, and so with several linked markers, their individual trait effects will be as shown in Fig.. Similar Arguments Apply to Other Populations Such as F_2s and RILs.

Table. QTL model for market analysis in doubles haploid (DH)lines from and F_1.(a)The F_1 constitution; M=Market and Q=QLT genotype.(b)DH genotypes,frequencies and genetic values.(c) Expertations of marker means and differences.R=recombination frequency between M and Q; m and a are

the mean and additive genetic effects of the QLT;δ is the additive genetic deviation at the market locus

(a)

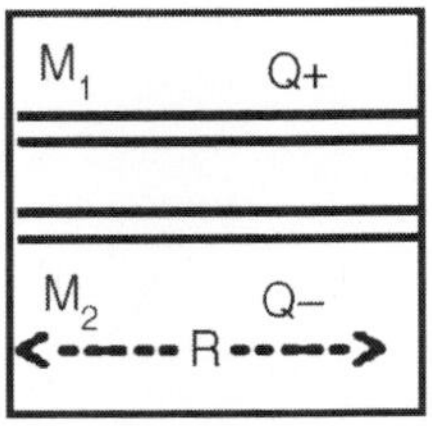

(b)

	F_1 gametes			
	$M_1 Q^+$	$M_1 Q^-$	$M_2 Q^+$	$M_2 Q^-$
Frequency	a(1–R)	a(R)	a(R)	a(1–R)
DH genotype	$M_1M_1 Q^+Q^+$	$M_1M_1 Q^-Q^-$	$M_2M_2 Q^+Q^+$	$M_2M_2 Q^-Q^-$
Genetic value	m + a	m – a	m + a	m – a

(c)

$$
\begin{aligned}
M_1M_1 &= \{a(1-R)(m+a)+a(R)(m-a)\}a \\
&= m+(1-R)a \\
M_2M_2 &= \{a(R)(m+a)+a(R)(1-R)(m-a)\}a \\
&= m-(1-R)a \\
\delta &= (\overline{M_1M_1}-\overline{M_2M_2})/2 \\
&= (1-2R)a
\end{aligned}
$$

If the trait means of the markers and their map are known very accurately, and only one QTL existed on the chromosome, then as Fig. shows, it would be relatively easy to locate the QTL. Figure shows the distribution of flowering time among a number of doubled haploid lines of *Brassica oleracea* containing alternative alleles at a marker located very close to a QTL. In practice neither the map positions of the markers nor their means are known with very great accuracy because the number of DH lines and replicated plots are generally few and the heritability of quantitative traits is generally low. Moreover, linked QTLs create difficulties because their individual effects can combine either to create the impression of an intermediate, ghost QTL, or to conceal their effects altogether. Various analytical procedures have been developed to maximize the efficiency and accuracy of locating individual QTLs and to separate linked QTLs. They all rely on the relationships described in the previous paragraph and use either maximum likelihood or weighted least squares regression procedures to estimate the QTL locations and effects that best fit the observed trait scores for each genotype.

The main difficulty is that there are two unknown parameters with respect to each marker score; the QTL effect, *a*, and map position. This means that it is necessary, in all methods, to use an iterative approach which involves trying all possible QTL locations along each chromosome and identifying the position or positions of the QTL which best fit the observed data as indicated by the size of the likelihood or the residual regression variance. As a result, a very large number of statistical tests are performed with the concomitant risk of false positive results.

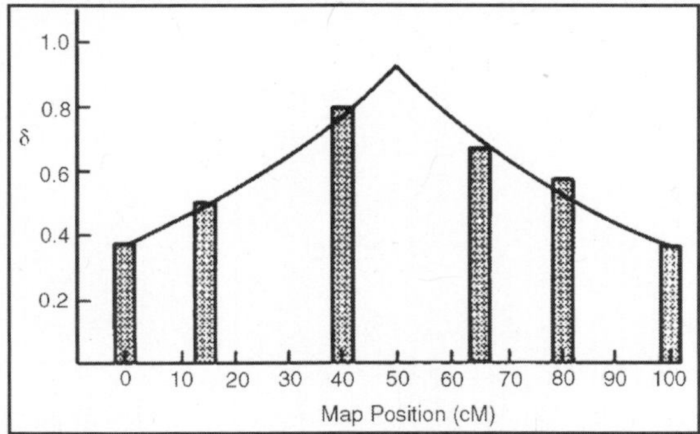

Fig. Quantitative Trait Effects (d) Associated with Marker Loci.The Bars Indicate the Observed Effects,While the Curve Illustrates the Expected Relationship if a QTL of d = 1.0 Existed at 50 cM

All methods provide estimates of QTL locations and effects with similar precision in terms of confidence intervals, and this precision drops rapidly as the heritability of individual QTLs declines, *i.e.* the estimates are accurate only when there are just two or three unlinked QTLs with large effects.

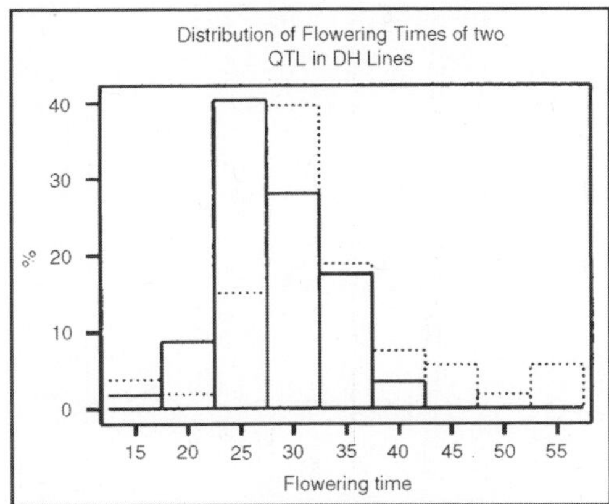

Fig. Distribution of Flowering Times Among Doubled Haploid (DH)Lines in Brassica Oleracea Containing alrernative Alleles of a Marker Linked to a QLT.The two sets Significantly Different Means Indicating a QLT with an Effect of Apporoximately ± 2 Days

For example, Hyne *et al.* (1995) showed that the 95% confidence interval for the location of a QTL contributing 10% to the phenotypic variance of an F2 could be as large as 35 cM. However, conservationists and breeders who are concerned with introducing QTL into commercial cultivars from other cultivars or more distantly related species, would normally only be interested in those cases where a few QTLs of major effect are to be manipulated. There is no evidence that dense maps are required for initial

QTL location. It is better to have large populations and a few (5–10) markers per chromosome.

There is evidence that many of the larger QTLs located so far map closely to previously known major genes. This is often used as evidence that there are, in fact, very few QTLs, but this may well be misleading for several reasons. The QTL map locations are sufficiently imprecise that they could well appear spuriously close to candidate loci. Only QTLs of large effect will stand much chance of being located and, by definition, they are likely to be major genes. It is likely that many cases where QTLs have very large effects could be due to the chance association of alleles of like effect at several QTLs along a chromosome, which cannot be separated as individual effects.

Much effort has been devoted to increasing the precision of QTL location, in particular in improving the power of the test to reduce the failure to detect genuine QTL. A long-established problem with breeding for quantitative traits in plants is the genotype 3 environment interaction. Using QTL location techniques, it is now possible to explore these effects at the level of individual QTLs although it is often difficult to know whether genotype 3 environment or poor repeatability is responsible.

ALIEN GENE TRANSFER

Once useful genes have been located on a framework map, it is then possible to use the molecular markers to facilitate the transfer of the useful genes between species and strains in an efficient manner. As stated earlier, recombination is not a frequent event along a chromosome and, furthermore, two recombination events rarely occur within a distance of 15 to 20 cM of each other.

It therefore follows that if a useful target gene (a major gene or a QTL) is known to be located within a region of chromosome flanked by two markers that are no more than 15 cM apart, then these markers can be used to follow and control the progression of the target gene through successive stages of a breeding programme. Clearly, the alleles at the marker loci have to differ in the donor and recipient cultivars, but any chromosome that contains A1 and B1 at the marker loci will also contain T1, the target allele to be transferred.

Should single recombination occur close to T, then the chromosome will contain A1 or B1 but not both. Providing the markers are less than 15 cM apart double recombination resulting in A1T2B1, and hence loss of T1, can be safely ignored. If the location of T is not accurately known, then three or more marker loci may be needed to be sure of safely bracketing the region containing T without fear of double recombination.There are conflicting requirements in marker-assisted gene transfer: the need to keep the bracketed region large enough to be sure of holding the gene to be transferred whilst not having it so large that too many other linked but undesirable alleles are transferred with it. Let us now consider the basic procedure of gene transfer.

Traditionally, breeders have introduced a useful allele into their cultivars by backcrossing the original F1 to the commercial cultivar whilst selecting at each generation for the desired allele. As was shown earlier, this can result in a very large region of chromosome around the target gene surviving even after ten generations of backcrossing – a phenomenon known as linkage drag. With markers, however, it is possible to reduce linkage drag considerably whilst simultaneously reducing the number of backcrossing generations to two or three instead of the normal six to eight. Moreover, the amount of plant material raised at each generation can be reduced because much of the molecular genotyping can be achieved at the juvenile stage.

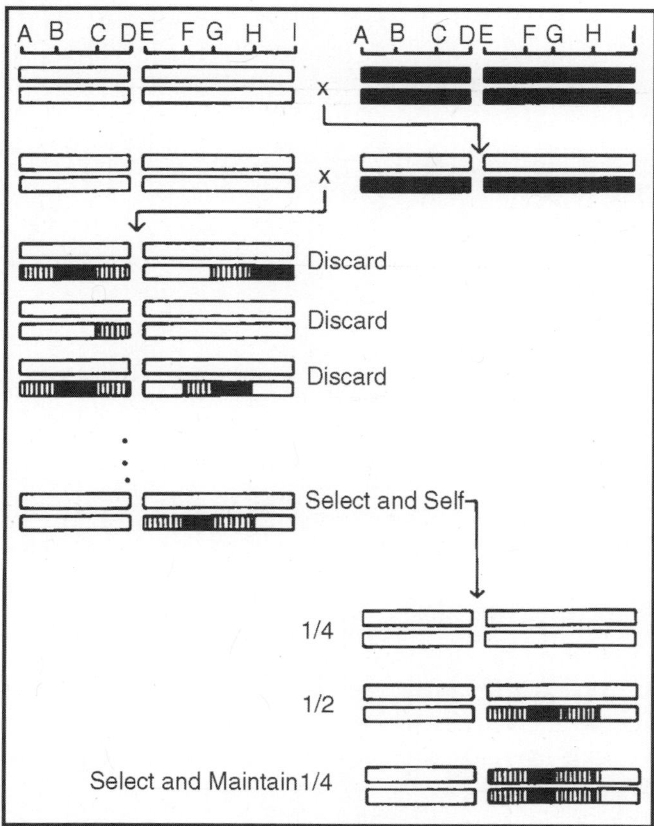

Fig. The use of Molecular Markers to Facilitate Gene Introgression.

The aim is to hold the target allele, T1, heterozygous throughout the backcrossing process, by selecting for the flanking markers, whilst simultaneously selecting for the genotype of the recurrent, commercial cultivar at all other loci. The latter can readily be achieved by having as few as four or five well-spaced markers on all chromosomes and selecting for recurrent parent alleles at each generation.

Depending on the chromosome number, some individuals in the first backcross generation will be homozygous for several whole chromosomes whilst still being heterozygous for the target gene. Those which have most of the recurrent parent chromosomes are selected and backcrossed again, and if necessary the process is repeated for a third generation. For each generation, fewer markers need to be screened as those found to be homozygous in the previous backcross no longer have to be checked.

Unless the position of the target gene is very accurately defined, it is prudent to have several markers covering its likely position, again with the proviso that they need to be separated by no more than 15 cM. At the final stage of the introgression process, a backcross individual which is homozygous for most of the recurrent parent alleles is selfed or, if this is not possible, intercrossed with a similar genotype. From among the progeny, individuals are chosen which have different combinations of markers bracketing the target region as shown in figure.

These can be selfed again and homozygotes for the different sequences identified and multiplied for reassessment for the target trait. Some will fail to show the target trait because of recombination, but of those that do, the lines with the shortest sequence of donor markers are selected for further trials. The approach can obviously be adapted to the simultaneous introgression of several genes: no new principles are involved but the screening process becomes more elaborate.

It is always possible at a later stage to reduce the length of the introgressed region around T by looking at other, more closely linked, markers. Similarly, it is wise to check the origins of the other chromosomes to avoid having introgressed unwanted regions through missing double recombinants in the central regions of individual chromosomes or single recombinants towards their ends.

If such errors are found they can easily be corrected by another round of backcrossing to the recurrent parent. When attempting to reduce the length of donor chromosome surrounding the target gene, it is important to realise that this should be done in two successive generations. If the sites of the required recombination events are less than 15 cM apart, then the simultaneous double event will not occur.

Even if they are far enough apart for the double event to occur, the probability of such an event may be too low to be practicable. If two recombination events each have a probability of 0.05 (*i.e.* 5%), their combined probability is at best 0.0025. Therefore, it would be necessary to genotype approximately 1200 individuals to be 95% sure of having at least one double recombinant. If it was tackled in two stages then it would require only the genotyping of 86 (= 28 in the first round plus 58 in the second; less being needed in the first round because the recombination could occur on either side of T), a very considerable saving in effort of 93%, or 1114 plants!

Marker-aided selection has been successfully tried for quantitative traits in several crops while the efficiency of the procedure has been examined by Lande and Thompson (1990).

MAP-BASED GENE CLONING

Providing a target gene can be located to within a pair of marker loci, it should be possible to use standard cloning techniques to walk along the chromosome between the markers and to find the target gene. The feasibility of this depends on the distance between the markers and the ease with which the target gene can be recognized from among the clones. Depending on the species, a genetic length of 1 cM can, on average, consist of as few as 280 kbp of DNA in rice, or 2220 kbp in maize. It is not always easy to locate a target gene with sufficient accuracy to be able to say that it is between two markers as close as 1 cM, and hence the actual lengths of DNA between them can often be much greater. With QTLs one can seldom achieve anything approaching this accuracy by existing methods of conventional mapping. Furthermore, these are average distances per centimorgan; some chromosomal regions could be much longer or, indeed, less.

Increases in mapping reliability can be obtained by using what are called near isogenic lines (NILs), which are lines obtained by selfing or backcrossing and are known to differ from some standard genotype by just a short defined section of chromosome. Providing this section is delineated by accurately mapped flanking markers, and the NILs which do and do not contain this region can clearly be shown to differ for the target gene, then the target gene has to be in that region. The genetic and physical proximity of the markers can be established by conventional mapping and by gene cloning.

Map-based gene cloning involves starting from one of the markers and walking to the next by means of a series of partially overlapping cosmid clones. Cosmid clones are necessary because they can be up to 50 kb in length and hence require fewer cloning steps in order to cover the DNA between the markers (280 to 2200 kb per cM in the examples cited above). Identifying the target gene is more difficult. Clearly, if the gene is between the two flanking markers it has to be on one of the cloned regions and could possibly be on two overlapping clones. It could be identified by transforming plants with each clone separately and identifying which transformant expresses the effect. If the gene concerned exhibits classical dominance it would be necessary to transform the homozygous recessive with the dominant allele or the transformant would not be recognized. Alternatively, the gene has to be introduced into a species that does not normally express the gene. Once the effective cloned fragment is identified, the actual location of the gene within the fragment could be identified by successively transforming sub-fractions of the clone or, following sequencing, by looking for potential open reading frames. In those cases where the target gene produces a known product, cDNA clones derived from tissue likely to be rich in the appropriate mRNA of the target gene can be used to identify likely sequences in the previous overlapping clones.

Once the gene has been identified by one of these methods, its structure and activity can be studied in detail. Disease resistance genes are obvious candidates for such studies as their action is very specific and their location on the genome of many species is well documented. However, unless suitable candidate loci.

GENE MAPPING

An important, though by no means an essential, step in genetic analysis is to produce genetic maps of the marker loci. Such maps are often referred to as 'framework maps' because they provide a framework within which important genes can be located, as well as providing a means of comparing chromosome organization in other closely or distantly related species.

There are two stages to mapping. Firstly, to arrange the markers in a linear sequence separated by an appropriate map distance, *i.e.* to construct a linkage map. The second is to relate the linkage maps to particular recognizable chromosomes. The latter is often the most difficult and generally not essential for breeding or conservation work. We will, therefore, concentrate on the former.

Chromosomes contain a single linear molecule of DNA and hence the markers on that chromosome occur at particular positions along that molecule. Typically, each chromosome contains 107 to 108 base pairs (bp), *i.e.* 104 to 105 kilo-base pairs (kbp) of DNA, while a typical structural gene, coding for a polypeptide chain, would be between 1 and 2 kbp long. Assuming only 10% of the genome is coding DNA, then a chromosome probably contains something of the order of 1000 to 10 000 genes. Fortunately, the fact that the chromosome is a linear molecule means that the genes need to be mapped in only one dimension.

Currently, the only useful method of gene mapping, at least as far as breeders and conservationists are concerned, relies on recombination between homologous chromosomes resulting from genetic exchange, which can be seen as chiasmata at diplotene through first metaphase of meiosis. A single chiasma on a particular chromosome results in half the gametes from such a meiosis being recombinant for that chromosome: half the gametes will contain a copy of that chromosome that is part maternal and part paternal in origin, while the remaining gametes will be entirely parental. Very few plant species have a sufficiently low number of clearly recognizable chromosomes that the number of chiasmata on a particular chromosome can be compared in different nuclei.

However, in many species the total number of chiasmata in each diplotene nucleus can be counted, at least in pollen meiosis, and the average number of chiasmata per nucleus calculated. The number and position of chiasmata on a particular pair of homologous chromosomes will vary from nucleus to nucleus; it is as though there are a large number of potential sites of exchange, but that in any given meiosis only a few sites are actually involved.

The longer the chromosome the more potential sites there might be. There will invariably be one chiasma and there may be two, three, four or more; a typical set of results is shown in Table 8.1a. A chromosome that has one chiasma on average is said to be 50 centimorgans (cM) long – a map length unit named after the American geneticist Thomas Hunt Morgan.

This relates to the fact mentioned above that one chiasma results in 50% recombinant chromosomes. By extension, a chromosome with an average of 2.5 chiasmata is said to be 125 cM (*i.e.* 2.5 350 cM) long. It appears that, as a rough rule of thumb, each chromosome has on average two chiasmata and so is 100 cM long. It follows, therefore, that providing the haploid chromosome number is known (n), the total map length will be approximately 2 3 n 3 50 cM. The average number of chiasmata per nucleus has been calculated in many species and this rule generally holds.

Table. Chiasma Fraquency and Mapping.Chiasma Frequency in Chromosomes of Secale Cereale.

Number of chiasmata per chromosome	Frequency	Percentage unrecombined Chromosomes
1	0.267	50.0
2	0.716	25.0
^{3}3	0.017	12.5
Mean	–	22.0

This rough rule is important because it gives the geneticist an idea of the total genetic map length that should be expected as more genes or markers are mapped. It provides a guide to the extent to which the currently mapped markers cover the full genetic map. Because the map is based in chiasma units, it is not the same as a physical map. The distribution of chiasmata is not uniform over a given chromosome in a species and will vary with the chromosome and the species.

Although a knowledge of the number of chromosomes and, ideally, the mean chiasma frequency per nucleus for the species, indicates the total map length, the actual map has to be constructed from examining the frequencies of progeny in crosses; *i.e.* maps are constructed from information based on the consequences of chiasmata, namely recombination of genetic markers. Recombination of genetic markers is identified by the presence of gametes that contain recombined markers, and such gametes can be recognized from the phenotypes of progeny of a cross. The frequency of such gametes is an estimate of the recombination frequency between the two markers concerned. The closer two markers are to each other on the chromosome, the less likely a chiasma will occur between them and the less likely they are to recombine. Providing the two markers are sufficiently close that either none or one chiasma occurs between them but never more, then the recombination frequency as a percentage is equal to the map distance in centimorgans as described above.

The problem arises when the markers are sufficiently far apart for two or more chiasmata to occur between them in some meioses. It can be shown that not only does a single chiasma between a pair of markers result in 50% recombination between them, but so also do two, three or more chiasmata, on average. It is necessary to say 'on average', because there are a variety of possible consequences with two, three or more chiasmata in any given meiosis but, in practice, when one is looking at the progeny of a cross or self, each progeny will be the result of gametes from different meioses.This fact implies that although map distance increases linearly with the number of chiasmata, the same is not true with the frequency of recombination, which can never be more than 50% for a pair of markers even though they might be 100 cM apart at opposite ends of a chromosome.

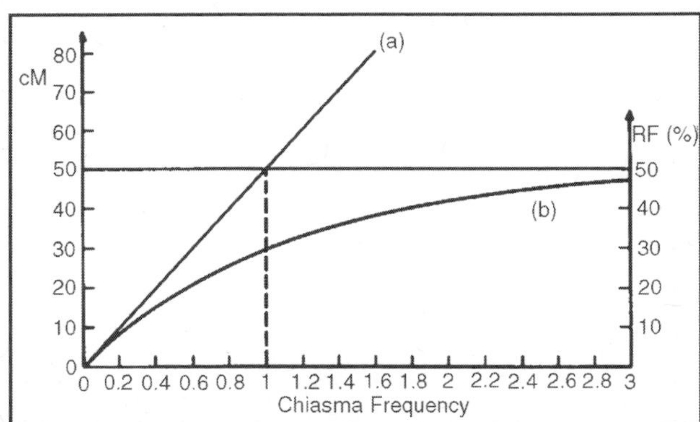

Fig. The Relationship of Chiasma Frequency With (a) Genetic Map Length and(b)Recombination Frequency.

These relationships are illustrated in Fig.. In order to overcome this problem, mapping functions have been devised to correct for multiple chiasmata. The most common of these are the Haldane (1919) and Kosambi (1943) mapping functions. Haldane's assumes that the probability of none, one, two or more chiasmata in a given interval follows a Poisson distribution, *i.e.* that the chiasmata are independent, random events. Kosambi's method allows for a certain degree of Non-independence in chiasma occurrences. It has long been known by cytogeneticists that chiasma interference occurs over quite large regions of a chromosome.

Interference means that if a chiasma occurs at a particular point on the chromosome, the next one will never occur closer than some fixed distance, the interference distance, from it. Beyond this distance, there is a short distance in which the interference disappears and the next chiasma can then occur randomly outside this range.

Observational data in various species of plants are confirming this and suggest that the interference distance is generally between 15 and 20 cM, *i.e.* 15 to 20% of a typical chromosome on either side of the chiasma. This has important and useful consequences for genetic and breeding work as will be shown later.Recombination frequencies between pairs of markers can be scored in a wide variety of different crosses. The simplest to use are generations derived from an F1 because only two alleles are segregating, and their distribution in the chromosomes of the parents of the F1 can be determined. The generations that can be used are F2, backcrosses (Bc), recombinant inbred lines (RILs) or doubled haploid (DH) lines, with F2s being the most informative.

General formulae for calculating recombination frequencies are given by Allard (1956). In outbreeding species where F1s are not available, a given individual can be considered as an F1 and its selfed progeny as being an F2, even though the inbred parents do not exist. If it cannot be selfed but controlled crossing to another individual is possible, then again those genes segregating in the cross can be mapped although the situation is more complex.

There could be from two to four alleles segregating at each locus and, with respect to any one locus, the cross could represent an F2, Bc1 or Bc2. Moreover, the distribution of alleles in the two parents cannot be known and has to be inferred from the progeny. This complexity is compounded by the large number of marker loci that may be segregating in any given cross.

Fortunately, a range of software packages are available to estimate these recombination frequencies, identify linkage groups, assign the markers to the most likely order and space them in map units (cM) on these linkage groups; examples are MAPMAKER and JOINMAP. Ideally the number of linkage groups should be equal to the chromosome number in the gametes but, unless the markers available provide a good coverage of the genome, it is likely that those markers on a given chromosome may appear as two or more separate linkage groups simply because the subsets are not sufficiently close for them to be recognized as being together on one chromosome.

A typical marker framework map is shown in Fig. for the chromosomes of *Brassica oleracea*. Clearly these maps provide a very detailed coverage of the chromosome. Many of the markers are very close, *i.e.* less than 10 cM, and over this distance it matters little which of the two mapping functions is used. With the more widely spaced markers, *i.e.* recombination frequency >15%, Haldane's function will exaggerate their distance apart because it will allow for more double crossovers than actually occur.

As more markers are mapped, the total length should converge on the value predicted by the chiasma frequency, *i.e.* approximately 2 3 n 3 50 cM. It is important to remember that genetic map distances have standard errors which are dependent on the size and type of population used to construct the map, the mapping population. The map obtained is that which best fits, given the data and the assumptions underlying it, but there may well be other maps which also fit the data well, though are slightly less likely.

Any particular map always develops a greater aura of respectability once it is published, frequently without stating its reliability, and subsequent yet different maps are treated with suspicion. A recombination frequency of true value p has a standard error of ?[p(1 2 p)/N], where N is the size of the gamete population. For example, if p is 0.1 (*i.e.* RF = 10%) and$N$ = 100, the standard error is 0.03 or 3%. In other words the 95% confidence interval of the estimate lies approximately between 4% and 16%.

There are other reasons, apart from statistical sampling, why maps could be wrong. The most obvious is the accuracy with which the data are scored and recorded. Autoradiographs and banding patterns on gels can easily be misread and research workers are loath to discard data even though its interpretation is ambiguous. All data should be checked independently by two or more people and any ambiguities removed. Wrong scores will bias the data and often exaggerate map lengths as they suggest double recombination; wherever double recombination appears to have occurred in two short, adjacent intervals, the data should be checked. Changes in methylation rather than scoring errors could cause a restriction site to appear or disappear so giving the false impression of double crossing- over, or it could simply be an error.

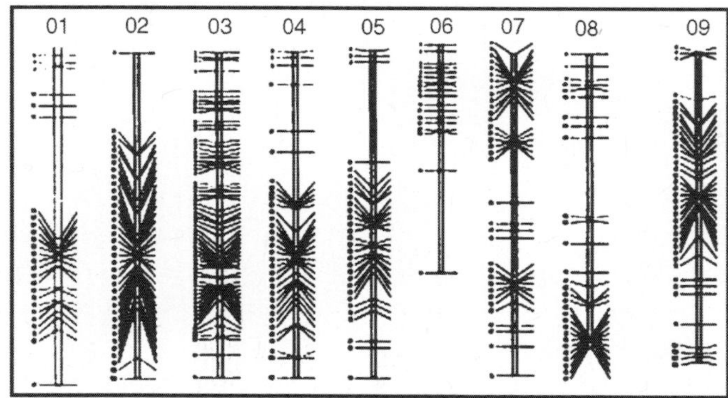

Fig. Typical Molecular Marker Framework Map.The Nine Groups of Brassica Oleracea Based on RFLP, Isozyme and Morphological Markers

Where maps appear to have major inconsistencies in different crosses, different chromosomal structural arrangements may be responsible, such as translocations or inversions. There is considerable evidence that chiasma frequencies, and hence recombination frequencies, may be quite different in male and female meioses even on the same plant although in other studies this may not be so. Thus although the order of markers on a linkage group may remain the same, the relative distances between them may be quite different if the map derives from male versus female meioses. For example, in *Brassica* it appears that the genetic map obtained from a backcross where the F1 is the female parent is 60% longer than when the F1 is the male parent. It is also clear that environmental factors during meiosis, particularly temperature, can affect the number and distribution of chiasmata, and this could be very critical if the crosses are set up at different times of the year.

Two or more different populations will almost certainly be segregating for different combinations of polymorphic markers, although some at least should be in common. These common ones can be used to overlay the two or more maps and a consensus map produced from an amalgamation of these. Again software is available to produce these consensus maps. Despite the various error-causing factors discussed above a considerable degree of consensus can be produced from such populations and, considerable similarity of chromosome sequence is conserved between even distantly related species, allowing the potential for cross-species genetic transfer.It was stated above that the total map length should be dictated by the chiasma frequency, in so far as this can be accurately measured. When the first few genes were initially put onto genetic maps in the first 40 years of this century, they only covered a small part of the genome – except in wellstudied species such as *Drosophila* and maize.

As more and more genes were placed on the map, so the total map lengths increased towards the asymptote dictated by the chiasma frequency. However, during the first decade following the use of molecular markers, the map lengths of some species appeared to exceed that expected, and it was even argued by some that the chiasma theory of recombination might be wrong.

No one would wish to argue that chiasma frequency data are free from error but there would have had to have been quite excessive underscoring to result in the disparity which was occasionally found. It would appear, however, that the excessive lengths were due in part to errors in scoring and to the use of small mapping populations, because subsequent map sizes have shown a progressive decrease towards the predicted asymptote. Scoring errors bias estimates of RF generally upwards, while RFs have to be large to be detectable in small populations. Our own experience with mapping in cereals and brassicas has also shown a progressive decrease in estimated map length as more markers can be found to fill some of the large gaps in the map and as errors are removed from the mapping data.

CONCLUSION

In order for genetic resources to be efficiently utilized in plant breeding programmes, it is first necessary to determine whether useful genetic variation exists in the material and secondly to develop the most cost-effective method of introducing the potentially useful genes into commercially acceptable material. Until recently, genetic analysis has relied on conventional segregation analysis in controlled crosses for qualitative traits determined by major genes, and on standard biometrical methods for quantitative traits controlled by polygenes. Now, however, the use of DNA techniques has opened up new possibilities for genetic analysis and breeding, and in this chapter the general principles and applications of these molecular marker-based procedures will be discussed.

In its specific sense, marker-assisted or marker-aided selection (MAS) is the use of appropriate easily recognizable genetic markers to facilitate or accelerate the selection of linked genes controlling useful agronomic traits. The nature of the marker loci themselves is not important and they are used simply to indicate the presence of useful alleles of genes of commercial or practical importance. However, the term has also come to include the wider use of markers to improve breeding programmes.

11

Fibrous Crop Residues: Factors Influencing Nutritive Value

TREATMENT OF STRAWS TO INCREASE DIGESTIBILITY

The productivity of animals fed straw can be in-creased by treatment of the straw to increase di-gestibility and feed intake, especially when combined with correct supplementation. The use of treated straw with supplements also allows the use of ani-mals with higher genetic merit. Methods for industrial-scale treatment of straw with alkalis (eg. ammonia and caustic soda) have been available for some time. In the developing countries the major problem with this approach is to find methods that are both acceptable and effective at the village level.

Table . Milk Production and Change in Liveweight (L Wt) in Zebu and Holstein/ Zebu cows in Bangladesh on Basal Diets of Untreated or Ammoniated Rice Straw (NH_3-straw) (Urea ensiled).The Native Animals per-formed best on the Untreated Straw, but the Crossbreeds were Superior when the Degradability of the Straw was Increased by Ammoniation.

	Straw	NH_3-straw
Straw intake (g/d)		
Local	–	–
Crossbred	5.2	10.2
LWt change (g/d)		
Local	–87	49
Crossbred	–211	168
	Milk yield (kg/98 days)	
Local	54	204
Crossbred	165	255

In some European countries, caustic soda has been used to increase the digestibility of straw for more than 40 years. A number of other techniques are be-ing applied in large enterprises (eg. ammoniation) but these methods have not been generally accepted by small farmers. Methods for treating straw with caustic soda are rarely economic, are difficult to ap-ply and are hazardous to people and animals. Thus, treating straw with caustic soda is impracticable in most developing countries.There are a number of ways in which ammonia can be used to increase the digestibility of fibrous feeds, including methods that use ammonia gas, am-monia in solution or ammonia generated from urea. Ammoniation of straw ap-pears to be potentially applicable to a wide range of situations.

The choice of the method of ammoniation will depend on the cost and availability of ammo-nia gas relative to urea. Ammonia gas lends itself to large operations where there is the necessary infras-tructure for distribution of ammonia in cylinders or tanks.

This may be readily done in oil-rich countries where ammonia gas is manufactured for agricultural purposes.

For small-scale farmers it is more convenient to gener-ate ammonia from urea by the "wet-ensiling process" . Urea is a common fertilizer which is often subsidised and which farmers have become accustomed to han-dling. It poses no hazard to human health but farm-ers are generally concerned about its possible toxicity to livestock.

Ammonia is generated rapidly from urea at high temperatures when it is mixed with moist stra w, which makes the system appropriate for trop-ical but not temperate countries.

A source of ure-ase may also need to be added for the more inert crop residues or fibrous byproducts (eg. bagasse; see Torres *et ai.* 1982). The ground seed from Jack-bean *(Canavalia ensiformis)* is being used commer-cially for this purpose in Colombia (Preston 1987).

Urea-ensiling appears to be less effective than using ammonia gas because of the formation of ammonium carbonate, which decreases the pH of the straw (Mason et *al. 1985).*Ammonia can be generated from other nitrogenous materials such as chicken manure but these tech-niques are only now being developed. The use of animal or human urine to ensile straw has been re-searched in Bangladesh and has shown promise for application on small farms. It is, however, essential to ensure an adequate level of urea in the urine.

Several other chemicals can be used to increase straw digestibility including calcium oxide, acids and acid gases (sulphur dioxide) or combinations of these. All these techniques are experimental and have not yet been proved to be applicable.One method of treatment to increase the digestibil-ity of fibrous feeds which appears to have some ap-plications on an industrial scale is the use of steam at high pressures.

This may be feasible where this energy source is available and inexpensive (ie. in sugar mills where there is usually surplus steam).

The technique appears to be particularly appropri-ate for treatment of bagasse which is available on site at the sugar mill and where the necessary technical knowledge and equipment are also available.

MINERAL CONTENT OF STRAW

The mineral content of straws is generally low and imbalanced but deficiencies are unlikely to be mani-fested in animals at maintenance or working. For pro-duction of meat and milk, requirements for minerals are increased many-fold and supplements should be supplied.

However, responses to mineral supplements will only occur after the major nutrient imbalances (protein and glucogenic energy) have been corrected.

The mineral composition of plants depends largely on the availability of minerals in the soil and closeness to the oceans (for N a). Calcium and phosphorus con-tents of straw are usually below recommended levels and cobalt, copper, sulphur and sodium may also be deficient.

The high concentrations of oxalates and silicates in rice straw also suggest that considerable amounts of calcium and magnesium salts can be lost as silicates and oxalates in urine and faeces.

DRAUGHT ANIMALS

Draught animals appear to work and survive on a wide range of fibrous diets and are able to tolerate the low digestibility of the diet and its low nitrogen content.

Working bullocks in Bangladesh con-sumed greater quantities of ammoniated rice straw (ensiled with urea) than their pair mates given un-treated straw but there were no apparent differences in work output or bodyweight change (Dolberg *et ai.* 1981).

Further evidence that working animals require little protein supplementation is the finding that young horses that were exercised grew at the same rate on a low-protein diet as those on a high-protein diet that were not exercised (Orton *et ai.* 1985a). Recent work suggests that excessive protein in a diet can slow the race horse (Glade 1983).

In Pakistan, Preston (personal observation) ob-served bullocks driving a press, crushing 400kg of sugar cane per hour, working alternate shifts of 3hours work followed by 3 hours for feeding and rest, and found that they were apparently able to work on this basis for 24 hours a day almostcontinuously for 6 months on a diet of only sugarcane tops.

In con-trast, a diet of sugarcane pith of higher digestibility (70%), supplemented with urea and minerals barely supported maintenance in 'growing' steers (Preston *et al.* 1976). The exclusive use of sugarcane tops as feed for working oxen iscommon practice on sugar es-tates with a priority for employing people. Obviously the influence of exercise on nutrient requirements is a subjectrequiring a great deal of research.

GROWING ANIMALS

Moderate rates of liveweight gain can be obtained with young ruminants on diets based on crop residues provided that a number of principles areapplied. These are discussed below.

Green Forage

A small quantity of highly digestible green forage ap-pears to have beneficial effects on rumen function on diets based on crop residues. The results in Table indicate that such improvements in the rumen ecosystem carry through to improved animal perfor-mance. *Azolla pinnata,* a water plant which grows symbiotically with the N-fixing alga *Anabaena azolla,* appeared to be even more effective than a mixed con-centrate meal in promotingliveweight gain in cattle fed a diet of wheat straw and sugarcane tops.

Table. Small amounts of the water plant Azolla pinnata appear to be as effective as a balanced concen-trate in improving the utilisation of a diet forcattle based on crop residues.

	Concentrate	Azolla
LWt gain (g/d)	140	330
Feed intake (kg DM/d)		
Wheat straw+sugarcane tops		
Supplement		
Total	3.6	3.0
Feed conversion(kg DM/kg gain)	26	9
OMD (per cent)	45	59

The advantage of using the foliages oflegume trees as "green" supplements is that these are rich in pro-tein (25 to 30% in dry matter), at least part of which appears to escape rumen fermentation. According to Bamualim *et al.* (1984)' leucaena leaf meal was as effective in stimulating voluntary intake of a low-N hay as was casein infused into the abomasum.The data in Table and Figure confirm the effectiveness of leucaena foliage as a supplement for cattle fed basal diets of rice straw. In the feedingtrial in Colombia, which involved 36 weaner steers fed *ad libitum* ammoniated rice straw for 90 days, fresh leucaena leaves (2kg/d) were as effectiveas 500g/d ohice polishings in stimulating liveweight gain (from 200g/d in unsupplemented animals to 550g/d with the supplements) .

Table. Leucaena Foliage (+LF) Increased Dry Matter Intake and Nitrogen Retention in Cattle given Diets Based on Untreated or Treated (sodium hydrox-ide) Rice Straw (NaOH-straw).

	Untreated straw		NaOH straw	
	–LF	+LF	–LF	+LF
DM intake, kg/d	5.1	6.6	3.6	3.9
DM digestion,	38	42	47	48
N retention, /d	–6	+7	–6	+26

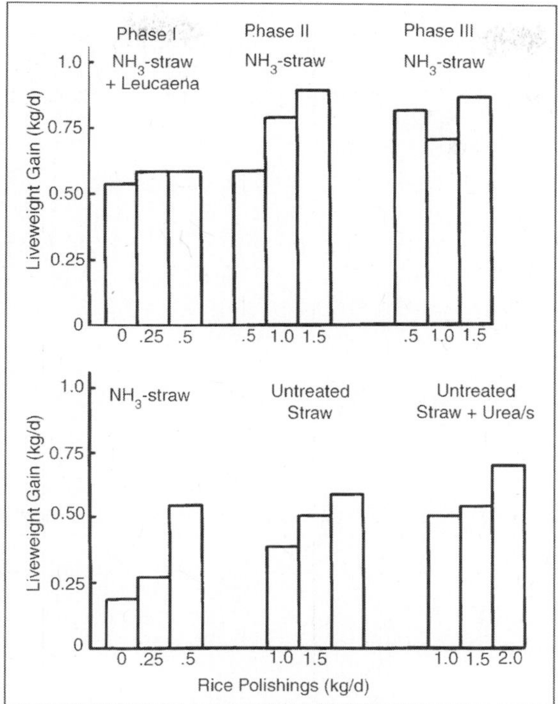

Fig. Development of a Cattle Feeding System Based on Ammoniated Rice Straw.

In phase 1, the basal diet of 6 groups of weaner bulls was ammo-niated straw (ammonia gas); the supplements were rice polishings at 0, 250 and 500g/d in the pres-ence or absence of fresh leucaena folzage (2kg/d). In phase 2, the treatments were ammoniated and un-treated rice straw with different levels of rice polishings. In phase 3, the treatments were ammoniated and untreated straw, the latter was supplemented with urea/S

Lucerne hay comprising 13% of a diet based on maize cobs increased liveweight gain of cattle by 100% when the basal diet was untreated and by 50% when the maize cobs were treated with ammonia.

Bypass nutrients and glucogenic compounds. The role of these nutrients in growing animals is dis-cussed in Chapter 4. The data in Table are from experiments in Australia with sheep fed oatstraw.

Table. Formaldehyde-casein (Formal-C), Rapeseed Meal (RSM) and Sunflower Seed Meal (SFM) (both Meals with and without Formaldehyde Treatment) Increased Fibre Digestibility, Dry Matter Intake, Liveweight Gain and Wool Growth in Sheep Fed a Basal Diet of Oat Straw.

	Urea	Formal-C		RSM	Formal-RSM	SFM
Formal-						**SFM**
DM intake (g/d)	1316	1779	1750	1899	1909	1977
Digestibility (per cent)						
OM	45	50	48	46	48	46

NDF	40	47	45	45	44	45
LWt change (g/d)	30	132	140	144	144	157
Wool growth (g/d)	4.5	11.0	10.5	10.1	10.1	11

The two feeding trials where the pattern of response to the supplement was measured. The results from the trial in Bangladesh are particularly interesting as they show that a daily supplement of only 50g of fish meal to cattle tripled productivity on a diet ofammoniated rice straw.

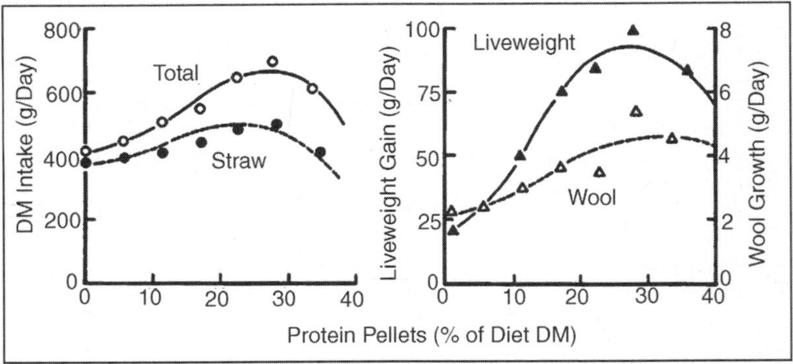

Fig. Feed Intake, Wool Growth and Liveweight Gain of Lambs given Barley Straw/urea and a Bypass Protein Pellet (Cottonseed Meal 80%, Meat Meal 8%, Soya Bean Meal 10%,Minerals 2%).

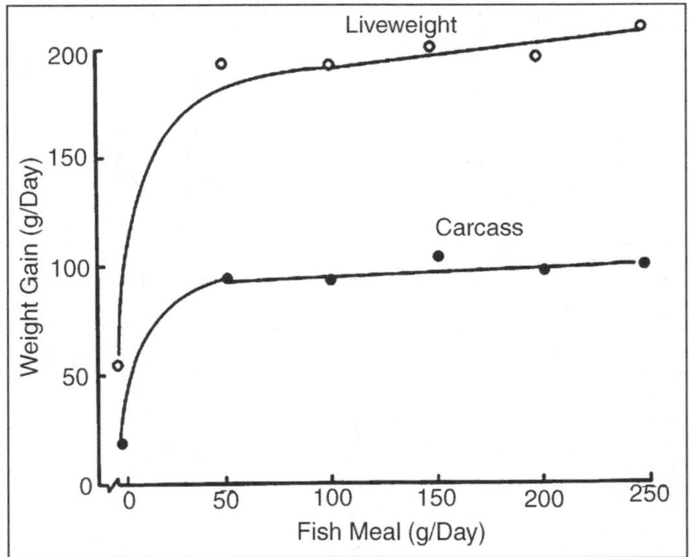

Fig. A Small Supplement of Fish Meal Dra-matically Increased g1'Owth in Live and Carcass Weight in Young Cattle Fed a Basal Diet of Ammoniated (urea-ensiled) Rice Straw in Bangladesh.

Recent research in Thailand and Australia provides strong support for the concept that the critical supplementary nutrients on a straw-based diet are bypass protein, starch and long-chain fatty acids. High rates of growth were obtained when the ammoniatedstraw (urea ensiling in Thailand and ammonia gas in Aus-tralia) was supplemented with starch, protein and oil in byproduct meals that are known to escape rumen fermentation .

Table. Effects of Supplementation of Ammoniated Rice Straw with a Mixture of Fat, Protein and Rice Starch. Young Brahman Bulls (36) Weighing150kg Liveweight were Used in an Experiment Lasting 153 Days.

	Supplement*, kg/d			1	2	3
Feed intake (kg/d)						
Rice straw	5.81	6.22	4.35			
Supplement	0.87	1.74	2.62			
Total	6.68	7.96	6.97			
LWt gain (kg/d)	0.47	0.84	0.93			
Feed conversion						
(kg DM/kg gain)	14	9	7			

Table. The Effects of Various Levels of By-pass-protein (Largely Cottonseed Meal) Supplement on the Live Weight Change of Cattle (320kg Liveweight) given a Diet of Ammonia-treated or Untreated Rice Straw, O.5kg Molasses/Block (15% Urea) to Provide Fermentable N, and O.6kg Rice Polishmgsto Supply Small Amounts of Starch and Lipid

Straw Preparation	Protein meal	LWt gain	(kg/d) (g/d)
None	0	38	
	0.4	365	
	0.8	292	
	1.2	306	
Treated with 3% NH_3gas	0	236	
	0.4	497	
	0.8	601	
	1.2	639	

The interaction between ammonia treatment of straw and response to bypass nutrients is illus-trated in Figure which summarises the perfor-mance of steers fed on ammoniated or untreated rice straw. During phase 2 (approximately 260-350kg liveweight), there were linear responses in growth rateto a supplement of rice polishings on both ammoni-ated and untreated straw. However, 2kg/d of the sup-plement was needed when untreated straw wasgiven compared with only 500g/d to support the same rate of liveweight gain on ammoniated straw. The inter-action continued to be manifested duringthe final finishing phase of the cattle (350-450kg liveweight), when urea and sulphur were sprinkled on the untreated straw.1t can be concluded that ammoniation raised the availability of nutrients from straw and also led to a more balanced array of nutrients to the cat-tle since lower levels of dietary bypass nutrients were needed to optimise performance.

METABOLIC DISORDERS ASSOCIATED WITH AMMONIATION OF FEEDS

Adding ammonia to ruminant feeds to provide fermentable N has been researched for a number of years. In early trials, ammoniation of molasses washighly detrimental because the ammonia gas reacted with the sugars, increasing the temperature and ap-parently resulting in formation of toxic methyl im-idazole compounds which induced a severe nervous disorder (Tillman *et al.* 1957, Bartlett and Broster 1958) .In countries with a temperate climate, ammonia-tion of straw (to increase digestibility) requires 3-6 weeks (versus 10 days or less in the tropics) because of low ambient temperatures. In order to reduce the treatment time, methods have been developed in which the straw is heated. Large ovensare manufac-tured in Europe to treat several tonnes of straw at 90°C, allowing the treatment time to be reduced to one day.

This practice has been commercialised in Europe and ammoniated straws have been fed to livestock with no reported ill-effects (H Sundstol, personal communication). However, ammoniated hays caused bovine hysteria when fed to cattle (for review see LaBore *et al.* 1984). The problem also occurs occasionally with ammoniated straw. This effect may be related to the proportion of the treated forage or straw in the animal's diet, since recent research in Australia (with one batch of rice straw and one of w heat straw) indicated that bovine hysteria devel-oped whenever the treated strawcomprised a large proportion (70-80%) of the diet. In studies by Perdok and Leng (1985, 1987), in which 64 yearling cattle were given thermo-ammoniated rice straw from a failed crop, almost all theanimals developed hyperexcitability, with the fol-lowing symptoms: animals were restless and blinked rapidly, their pupils dilated and their vision was ap-parently impaired; involuntary twitching, trembling, loss of balance and frequent urination and defeca-tion were observed. In addition, respiration rate was high, heart rate was low and the animals salivated copiously; they bellowed and perspired. The most obvious and most dangerous symptom wassudden stampeding: the cattle galloped in circles, were in-clined to collide with each other and also to run into fences. Often the animals galloped at suchspeed that they broke limbs, and in one instance an animal died. The symptoms usually lasted for 5 minutes and were repeated at 20- to 30-minuteintervals. The affected animals appeared normal between these attacks and tended to return to feed on the treated straw. The condition was alsoinduced in animals fed the same rice straw after it had been treated with ammonia under plastic sheets for 4 weeks.

The syndrome has occurred also in young calves suckled by cows consuming a diet containing a ma-jor proportion of ammoniated rice straw: both cows and calves showed symptoms. Pasteurised milk from these cows also caused the hysteria when fed to calves. Under some circumstances calves suckled by cows fed ammoniated hays have died. These obser-vations indicate that the toxic compound(s) is trans-mitted in milk and is unaffected bypasteurisation Bovine hysteria in cattle fed straw has not so far been reported from Northern Europe, possibly be-cause the proportion of ammoniated straw fed rarely exceeds 30% of the total diet in intensive livestock feeding systems. A recent report from Southern Spain (Cabrera *et al.* (1987) has, however, shown that in warm climates feeding ammoniated straw to cattle and sheep can be extremely dangerous. Cabr-era *et al.* (1987) reported numerouscases of hyper-excitability and death in sheep and cattle fed straw when this was initially treated with ammonia gas af-ter 1100 hours and on a warm day in summer.

Ammoniation through urea ensiling is highly un-likely to result in toxic compounds being produced, particularly if the moisture content of the straw iskept high. It is advisable, however, always to pre-pare straw when the temperature is low.The transmission of toxic compounds in milk from cows fed ammoniated forages suggests that care should be taken in feeding ammoniated forages tomilking animals.

LONG-CHAIN FATTY ACIDS

Evidence which demonstrates the benefits of having a dietary source oflong chain fatty acids in straw-based diets is summarised in Figure. Lambs fed ammoniated wheat straw and minerals increased their growth rate and feed conversion in response to di-etary long chain fatty acids given in the form of in-soluble calcium soaps (Ca-LCFA). The response was only apparent when bypass protein was also given. Subsequent work confirmed theinteraction between LCFA and bypass protein, and showed significant increases in carcass fatness in response to the LCFA supplement (given as long-chain fatty acid prills).

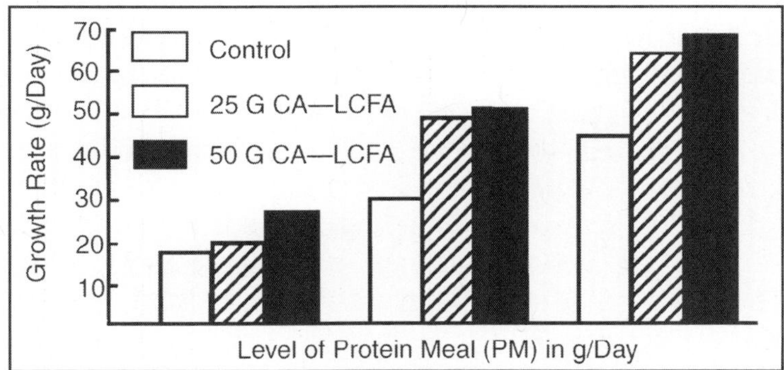

Fig. A Supplement of Long-chainfatiy Acids (as Insoluble Calcium Soaps) Increased the Growth Rate of Lambs Fed Ammoniated Wheat Straw and Bypass Protein

From the discussion in earlier chapters it is obvious that fibrous crop residues can only support low milk yields in ruminants because of their inability tosup-ply enough protein and glucogenic energy to balance the VFA energy. These feeds are the most, and often the only, available resource on manysmall farms ill developing countries. Therefore every attempt should be made to use them as efficiently as possible. Milk synthesis represents a greater drain on critical nutri-ents than any other physiological state. In view of the nutritional limitations of most crop residues, set by low digestibility and low N content, and the high demand for amino acids, long-chainfatty acids and glucogenic compounds, milking an-imals must be given high priority for:

- The available supplements and
- The residues that have higher potential digestibility or that have been treated to improve digestibility.

The scientific basis for feeding milking animals on crop residues is discussed in the following section.

Effect of Ammoniation

Upgrading the nutritional value of straws by ammoniation (urea ensiling) is more likely to be economi-cally justifiable for lactating animals than foranimals in other physiological states. In most circumstances the sale value of the extra milk produced should more than cover the cost of processing thestraw.

Table. Milk Yields and Changes in Live Weight (L Wt) in Cattle and Buffaloes Fed Basal Diets of Un-treated and Ammoniated (Urea-ensiled) Rice Straw (NH_3-straw)

Straw	Milk yield (kg/d) NH_3-straw		LWt change (g/d) Straw NH_3-straw	
Cow[1]	1.1	2.3	–149	+109
Cow[2]	2.4 (4.6)	3.4 (4.9)	–266	+93
Buffalo[3]	2.4 (6.8)	3.2(7.6)	–17	+93

The data in Table summarise research from Bangladesh and Sri Lanka where ammoniated or un-treated rice straw was the basis of the diet for Zebuand buffalo cows. The cattle responded significantly to ammoniation of the straw, both in milk yield and in liveweight change. Ammoniation of the straw in-creased feed intake, and obviously reduced the need for bypass nutrients since the amount of milk produced per unit of concentrate fed wasincreased by an average of 53%. The proportion of the total diet represented by concentrates was reduced from an av-erage of 14% to 9% due to increased intake of the treated straw.

Table. Ammoniation of Rice Straw (NH_3-straw) permits Reductionin Amount of Concentrate Required for Milk Production

	Straw	NH_3 straw
Concentrate (per cent diet DM)		
Zebu x Friesian	11	8
Zebu	21	14
Buffalo	11	5
Milk produced (kg/kg concentrate)		
Zebu x Friesian	1.7	2.7
Zebu	2.8	4.2
Buffalo	4.0	6.0

Effect of Green Forage

The effect of green forage, in the form of gliricidia leaves (*Gliricidia sepium),* is illustrated in Figure. On a diet of untreated straw, feeding gliricidia foliage at approximately

15% of the dietary dry matter in-creased milk yield by 22%. With ammoniated straw as the basal diet gliricidia comprised 10%of the diet and milk yield was increased by 14%.

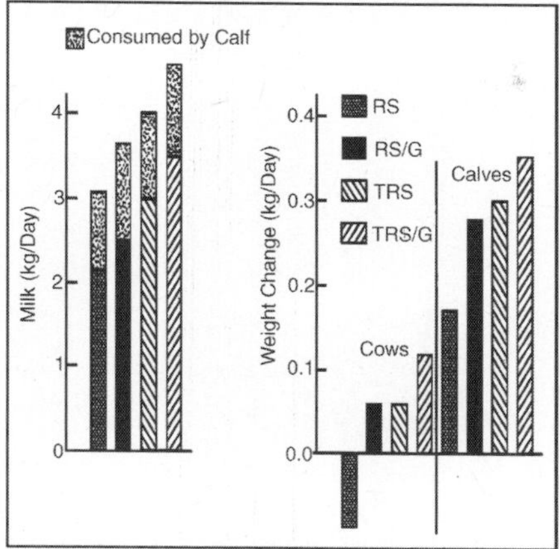

Fig. Supplementation with Fresh Gliricidia foHage (G) or Ammoniation of Straw (Urea Ensiling) (TRS)increased Milk Yield and Body Weight Gain of Buffalo Cows and Calves Fed a Basal Diet of Rice Straw and 1 kg/day of Concentrate (RS). Best Results were Obtained when Straw was Treated with Ammonia and Supplemented with Gliricidia Goliage (TRS/G)

Bypass Nutrients

The data in Figure and illustrate the results of trials in which responses different amounts of protein supplements were measured. In Bangladesh there was a linear increase in milk yield in Zebu cows when fish meal was given as a supple-ment to a basal diet of ammoniated rice straw.Milk yield was increased by 23%, fat percentage by 8% and liveweight gain by 110% when 1000g of coconut cake were fed daily to lactating buffaloes in Sri Lanka on a basal diet of ammoniated rice straw and minerals.

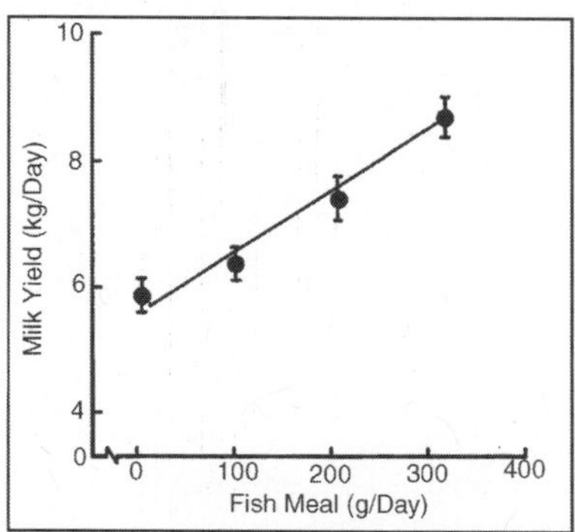

Fig. Milk Yield of Native and Crossbred Cattle Fed Ammoniated (Urea-ensiled) Rice Straw in Bangladesh was Increased Linearly with. Small Amounts of Fish. Meal (0 to 400g/d).

Table. Supplementing Ammoniated (Ureaen-Siled) Rice Straw with. Coconut Cake Increasedthe Milk Yield and Liveweight Gain of Buffaloes inS7'i Lanka.

	Coconut cake, kg/d		**0**	**1**
Milk yield (kg/d)	2.6	3.2		
Fat (per cent)	9.1	9.8		
LW gain (kg/d)	0.1	0.21		

WOOL GROWTH

The effects of strategic supplementation on wool growth in sheep are illustrated in Figure. Wool growth was increased when either a protein meal(cottonseed cake) or a readily-digestible forage (lucerne hay) was the supplement in a wheat-straw diet pro-vided with fermentable N by adding urea or by ammoniation of the straw.

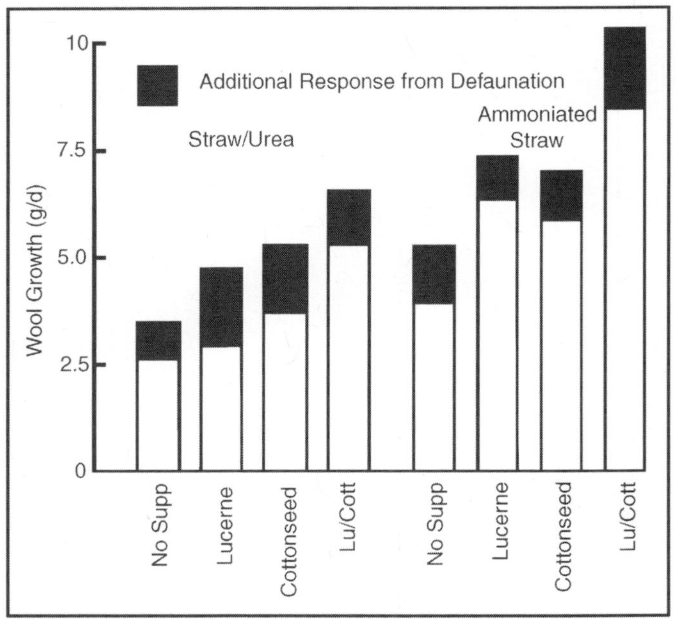

Fig. Supplements that Provided Rumen "Acti-vators" (Lucerne Hay) and/orbypass Protein (Cotton-seed meal) Increased the Wool g1'Owth of Sheep Fed a Basaldiet of Wheat Straw that had been Sprayed with Urea or Ammoniated.

There wereadditional benefits when the sheep were defaunated (Source: S H Bird, B Romuloand R A Leng, unpublished data.) The combination of the two supplements gave the best results. Responses were more marked when the straw digestibility was increased byammoniation and when the sheep were defaunated.

The data presented previously suggest that the optimal wool growth response in growing lambs fed a barley straw/urea basal diet occurredwhen the protein meal (80% cottonseed meal, 10% soya bean meal, 8% meat meal and 2% minerals) comprised about 30% of the diet dry matter.

FIBROUS AGRO-INDUSTRIAL BYPRODUCTS

There are a number of fibrous residues that, un-treated, have only limited application as the basal component in livestock feeds (eg. bagasse, palm-pressed fibre, cocoa pods and rice hulls). The lim-itation to all these feed resources is their extremely low digestibility and, even when supplemented with essential nutrients, intake is too low to support main-tenance. They are, however, often used as fillers and sources of roughage in high-concentrate diets.

Most of these residues arise from industrial process-ing and are therefore available in large quantities at the factory site. The concentration of these byprod-ucts at the factories is an incentive to finding ways to use them as ruminant feeds by using 'relatively so-phisticated' technology to increase their digestibility. Techniques such as briquetting, steam treatment and ammoniation can be considered under these circum-stances.

The factories where these residues are produced usually have the infrastructure necessary to enable adequate servicing of the machinery that is normally required for any large-scale industrial treatment of a fibrous residue. Often the residue is used as fuel in the factory (eg. bagasse, palm-pressed fibre and peanut hulls), but frequently there are surpluses. Little re-search has been done on these residues, with the ex-ception of steam-treated bagasse which is now being used commercially in diets for cattle in both Brazil (E L Caielli, personal communication) and Colombia (Preston 1987).

Some of the original research on steam treatment of sugarcane bagasse was done in Mauritius (Wong et al. 1974). Treating the bagasse with high pressure (14kg/cm2) steam for 5 minutes raised dry matter di-gestibility from 28 to 60% (rumen nylon bag method; 48 hr incubation). Early attempts to use the treated bagasse as the basis of the diet for growing cattle focussed attention on the need to supplement with bypass protein (fish meal), LCFA and glucogenic pre-cursors (maize grain).

Table : Calves lose weight on a diet of steam-hydrolysed bagasse (200°C for 10minutes) supplemented with only urea and minerals. Significant improvements in growth and feed conversion wr'e brought about when fish meal (bypass protein) and/or maize grain (glucogenic energy and oil) were added to the diet.

(0.25kg	**Control 1**	**Fish meal (0.25kg/d)**	**Maize meal (1kg/d)**	**Fish meal + maize +1kg/d)**
Bagasse intake(kg DM/d)	4.0	4.4	4.5	4.5
LW change (kg/d)	–0.17	+0.08	+0.16	+0.33
Feed conversion (kg DM/kg gain)	–	34	23	12

The data in Figure show how the commercial feeding system was developed in Colombia, starting from the premise that the treated bagasse should contribute the major part of the diet and that locally available supplements should be used. The diet which gave the best results (810g/d of liveweight gain), and which was chosen for the commercial programme, contained (per cent dry matter basis): 53 steamed bagasse, 16 final molasses, 2 urea, 15 gliricidia foliage, 6 poultry litter, 7 rice polishings and 1 salt.

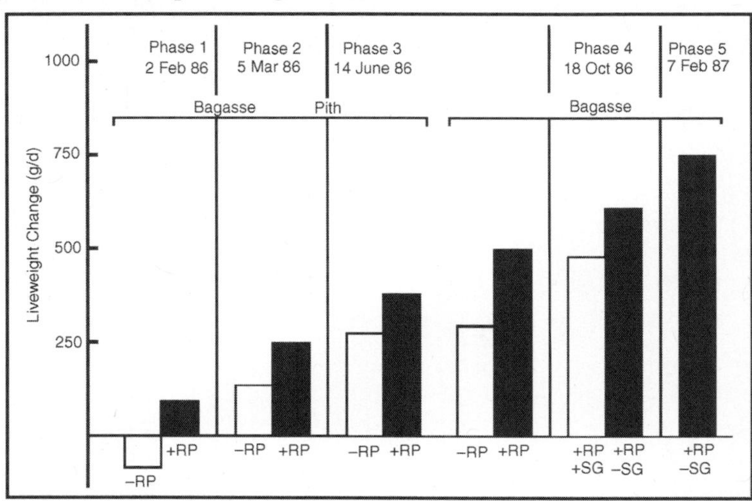

Fig. Development of a Cattle Feeding System Based on Hydrolysed Cane Bagasse (Steam at $14kg/cm^2$ for 5 Minutes).

The same 8 groups of animals (weaned bulls of 180kg initial liveweight) were used throughout the twelve month trial. In phases 1 and 2, the basal diet wastreated bagasse pith supplemented daily with molasses/urea (10% urea), gliricidia foliage (2%of liveweight daily) and poultry litter (0.2% of liveweight). The variable was the presence or absence of rice polishings (0.2% of liveweight, RP).

In phase 3 the treatments were untreated bagasse and bagasse pith, with or without the rice polishings. In phase 4 the effect of additional long fibre was studied (from African Star grass S G) with the basal diet using treated bagasse and with the addition of rice polishings.

In phase 5, all the animals were given the best diet based on the results obtained in the previous phases

The molasses was used as a carrier for urea and also provided trace elements; gliricidia foliage was the source of fibre, vitamin A and some fermentable and bypass protein; poultry litter provided fermentable N and macro (Ca, P, S) and micro (Cu, Co, Zn etc) minerals and rice polishings was the sourceof nutri-ents which improved the balance of amino acids and glucose available in the total nutrients absorbed and this increased the efficiency of feed utilisation. The lipid content of rice polishings was expected to be used highly efficiently in this otherwise low fat diet.

FRESHLY HARVESTED GRASSES

Elephant (or Napier or King) grass *(Pennisetum pur-pureum),* guinea grass *(Panicum maximum)* and sugarcane *(Saccharum officina rum*) are among the high-est yielding perennial crops, in terms of total biomass production per unit area and efficiency of solar en-ergy capture. *Pennisetum*species and sugarcane are used widely in the American tropics as a reserve crop ("ensilaje vivo") for feeding during the dry season. Sugarcane has the advantage that its energy value as feed increases as it matures, since the accumulation of sucrose more than compensates for the increasing lignificationof the cell wall. The "stress" of the dry season (low soil moisture, low soil N and [generally] lower air temperatures) also stimulates the storage of sucrose in the stem.

Pennisetum yields most biomass when it is har-vested at 6-month intervals, although it is well understood that nutritive value is highest when harvesting is over shorter intervals (about 6 weeks). On the farm, it is difficult to maintain the rigorous practice of fertilizer application, irrigation and frequent cutting and as a result the N content and dry matter digestibility of the infrequently harvested for-age are low.

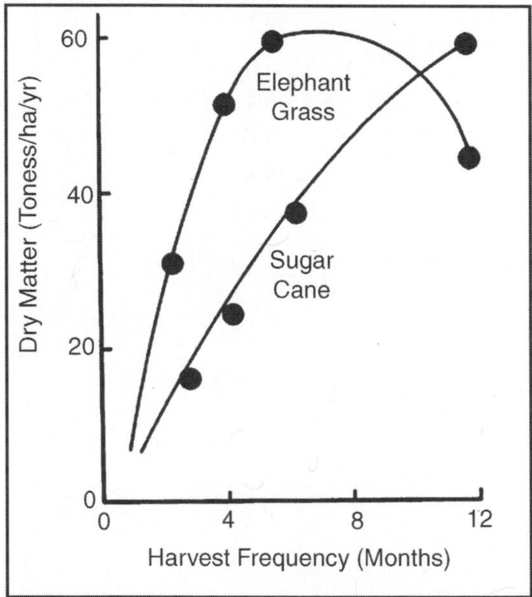

Fig. Annual Biomass Yields from Sugar Cane and Elephant Grass Harvested at Different Intervals

Early work with these forages emphasised their use as "roughage" supplements in concentrate diets for intensively managed (usually confined) milking and fattening cattle and buffaloes. This policy predicated against the efficient use of these feed resources since the drop in rumen pH when cereal grain is fed depresses the digestion of fibre.

In the following section examples are given of re-cent research in which these forages have been the basis of the diet and have been strategically supple-mented.

Legume Forages

In Costa Rica , lactating goats were fed a basal diet of King grass (*Pennisetum purpureum*) supplemented with a fixed allowance of green banana fruit and in-creasing amounts of the leaves of the legume tree *Erythrina poeppigiana*. Total dry matter intake and milk production increased linearly with legume sup-plementation. There was only a minimal substitution of the King grass in the diet (intake fell from 690 to 600g dry matter / day).

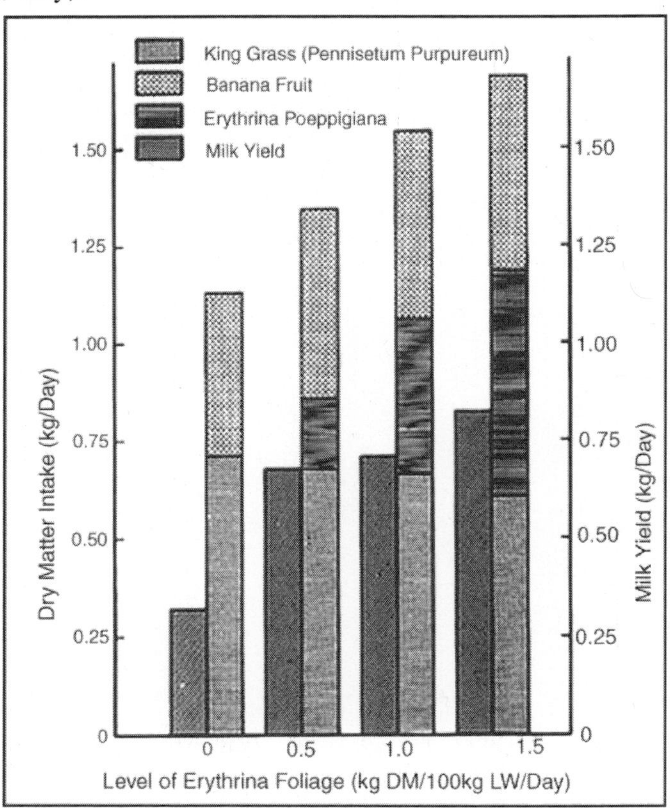

Fig. Effect on Dry Matter Intake and Milk Yield of Goats of Supplementing their Basal Diet of King Grass and Banana Fruit with Foliage of Erythrina poep-pigiana.

In research in Colombia the foliage of the legume tree, *Gliricidia sepium*, was given as a supplement to weaned steers fed a basal diet of freshly har-vested King grass during the dry season. Growth rate increased curvilinearly in response to in-creasing levels of the legume foliage, with the opti-mum legume content of the diet being about 30%.

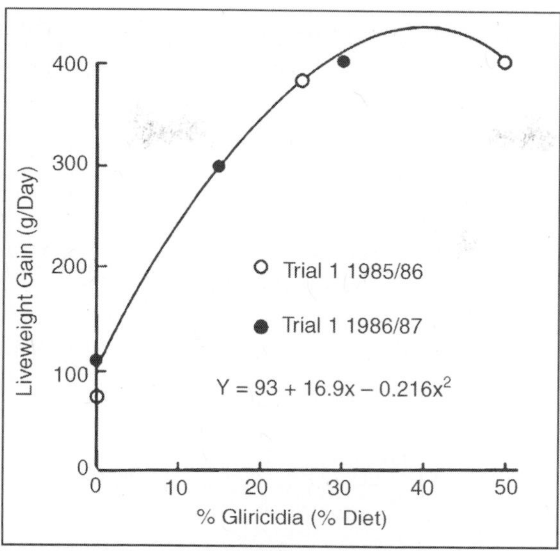

Fig. Effect of Increasing Levels of Gliricidia Foliage on the Growth Rate of Weaned Bulls (Initial Weight 180kg; 4 Animals per group) given a Basal Diet of King Grass

Similar effects have been reported with West African Dwarf goats in Nigeria fed combinations of Guinea grass (*Panicum maximum*) and gliricidia (M. van Houtert, personal communication).

Table. Effect of Supplements of Leaves from Gli-ricidia Maculata on Performance of Glowing Sheep Fed a Basal Diet of Brachiaria Mileformis.

Feeding a supplement of gliricidia to growing sheep on a basal diet of freshly harvested *Brachiaria multi-formis* gave a significant response inbodyweight gain and in wool growth when the legume comprised up to 28% of the diet.

Table. Effect of Supplements of Leaves from *Gli*-Ricidiamaculata on Performance of Growing Sheep Fed a basal diet of Brachiaria Mileformis.

	per cent Gliricidia in diet (DM basis)			
	0	28	50	65
LW gain, g/d	25	43	44	39
DM intake, g/d	530	610	690	690
DM conversion	21	14	16	18
Carcass, kg	10	12	12	12
Clean fleece, g	280	350	320	360
Staple length, cm	4.3	4.7	5.4	5.5

Results of studies with milking cows fed diets based on freshly harvested Setaria grass and given a range of supplements are shown in Table Themost appropriate supplement was groundnut cake since it maintained body weight and supported the same lev-els of milk production in cows as cerealgrain or molasses-based supplements fed at twice the rate. The implication is that the principal deficiency in the absorbed nutrients was amino acids and this was effectively corrected by the ground nut supplement, which provided bypass protein.

Table: Effects of supplements based on molasses, cereal brans/oil cake or groundnut cake onmilk yield and liveweight change of Friesian cows given a basal diet of freshly harvested grass (*Setaria kazangula*) ad libi-tum.

	No suppl.	Cereal/oil cake	Molasses/urea cake	Groundnut
Milk yield, kg/d*	6.1	7.6	6.8	7.9
LW change, kg/d	–0.72	0.0	–0.83	0.13

SISAL BAGASSE AND PULP

INTRODUCTION

Sisal *(Agave fourcroydes)* is a source of fibre for twine, ropes and carpets. The entire leaf blade is harvested and transported to the factory, and there are thus no residues in the field. However, the bagasse that remains after the fibres are extracted is potentially useful. The bagasse left after traditionalprocessing contains both long and short fibres. More advanced factories extract a greater proportion of the fibres and some of the saponins, which areused as raw ma-terial for the manufacture of steroid hormones. The byproduct of this modified process is referred to as "pulp" to differentiate it frombagasse (Riley 1984).

Data on the chemical composition of both these byproducts are given in Table. Because of the high content of soluble sugars and the high moisture content, the pulp or bagasse begins to ferment imme-diately after fibre extraction and most of the soluble sugars are converted to lactic acid.

Table. The Chemical Composition of Sisal Bagasse and Pulp

	Sisal pulp		Sisal bagasse		
	Fresh	Ensiled	Fresh	Ensiled	Fresh
Proximate analysis (per cent DM)					
Crude fibre	29	30	28	29	23
Crude protein	5.3	4.9	5.3	5.5	6.3
Ether extract	3.1	3.0	3.7	3.4	3.0
Ash	13	13	15	15	12
Water-soluble CHO	21	0	27	0	
Mac1'O-minerals (g/kg DM)					
Calcium	51	53	47	49	35
Phosphorus	1.1	1.0	1.0	1.2	1.5
Magnesium	8.0	8.0	9.0	10.0	8
Trace minerals (mg/kg DM)					
Zinc	1.2	1.2	1.5	1.4	
Copper	0.5	0.5	1.0	0.8	
Cobalt	ND	ND	ND	ND	
Manganese	1.0	1.1	1.0	0.8	
Iron	2.0	1.8	2.3	2.2	
0rganic acids (per cent DM)					
Lactic	1.6	16.5	1.0	18.1	
Citric	1.0	0.5	1.2	0.8	
Oxali,	5.5	5.3	5.2	5.4	
Ph	4.0	3.9	3.9	3.8	

The constraints associated with the use of the pulp/bagasse as a feed for ruminants relate partly to the content of organic acids, principally lactic andoxalic acids. Phosphorus and some of the trace ele-ments are also severely deficient, and thus the min-erai content of pulp or bagasse is highlyimbalanced relative to the needs of productive animals.

Cattle fed only on ensiled sisal pulp for a long pe-riod developed acidosis and barely maintained body-weight, despite the reasonably high digestibility(50-60%) of the feed . Attempts to correct only the acidosis and the min-eral imbalance did not increase animalproductiv-ity (Naseeven and Harrison 1981; Belmar and Riley 1984) .Research in Mexico on the use of sisal pulp as a feed for ruminants (Rodriguez 1983) is a further ex-ample of the application of the principles developed in this book, namely that attention should first be given to optimising the rumen ecosystem and then to correcting the protein/energy ratio with bypass pro-tein. Only after these two steps have been taken are responses to supplementary minerals observed.

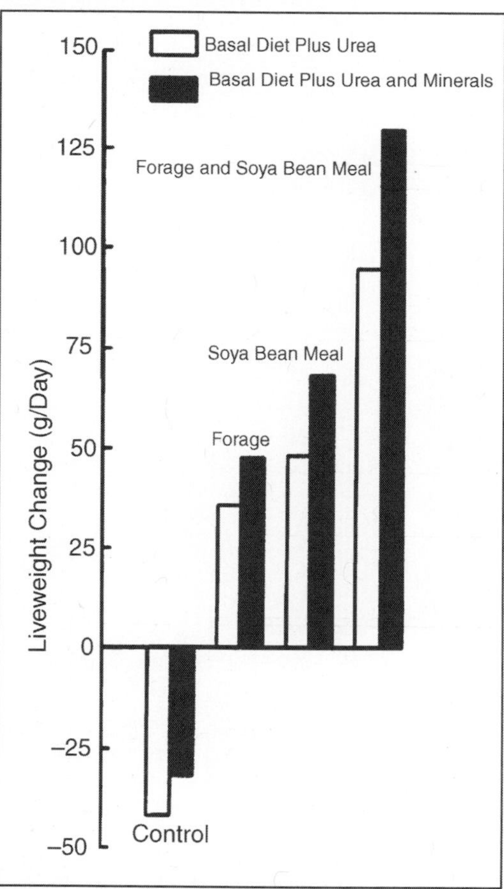

Fig. Sisal Pulp is Relatively Digestible (55-60%) but Low in Fennentable-N, Protein and Minerals.

Supplementing with urea and mineTals had no effect on lambperformance until green forage (foliage from the tree *Brosium alicastrum*) and/orbypass protein (soybean meal) were also given

BANANAS AND PLANTAINS

Bananas and plantains provide voluminous amounts of crop residues which are normally incorporated into the soil for mulching and to maintain soil fertility. When bananas are grown for export, up to 20% of the fruit may be rejected, and these are often discarded.

There are few nutritional constraints to the use of banana fruit as feed for ruminants. However, it is important to feed the material while it is green andthe carbohydrate is still in the form of starch. The fruit can be fed as a complete bunch (or "hand") and there is no need for any processing. The fruit islow in nitrogen and the skin is rich in tannins. The greatest nutrient deficiency of bananas as a feed for ruminants is fermentable nitrogen and dietscontain-ing bananas must be supplemented with urea.

Where molasses is freely available, the most con-venient procedure for supplementing cattle with urea is to allow the cattle free access to aurea/ molasses mixture (10% urea). Cattle will restrict their intake of this supplement to about 2kg/day, which provides enough urea (200g/day) to ensure an adequate level of ammonia in the rumen for efficient fermentation (McEvoy and Preston 1976). If a high-protein forage is fed with the diet,either as a legume component of a pasture in a restricted-grazing system or as a protein-rich foliage (eg. leucaena or gliricidia), there appears to be no advantage from supplementing with a bypass-protein meal. Results from feeding reject bananas to cattle in Colombia are summarised in Table.

Table. Growths Rates of Cattle Fattened Under Commercial Feedlot Conditions in Colombia on Reject Bananas (40% of Diet DM), Chopped Elephant Grass (42%) and Kudzu Legume Forage (18%) with Different Supplements.

Protein suppl.	Initial LW (kg)	Daily gain (kg/d)	Days on feed	Comparisons
No supp!.	330	0.83	90	Female adult
	325	0.64	90	Male adult
	334	0.80	90	Female old
	325	0.88	90	Female young
	240	0.91	100	Male young
	350	0.61	100	Male old
1kg/d CSC	381	1.15	60	+ sulphur
	391	0.97	60	– sulphur
	332	0.98	60	Ureal Jllolasses
	322	0.97	60	Aqueous urea
2kg/d CSC		1.60		Chianina×Zebu
		1.14		Zebu crosses
		1.25		Zebu bulls

	1.10	Zebu steers
	1.66	Chianina bulls
	1.50	Chianina steers

Banana Foliage

The dry matter in the leaves and pseudostem of the banana plant is relatively digestible (65 and 75%, respectively; Ffoulkes and Preston 1978a).Despite this, these feeds barely support maintenance if given alone to ruminants. However, when ruminants are fed banana pseudostem they respond to supplementation with urea, highly digestible green forage and bypass nutrients. The data in Figure show that cat-tle increased their voluntary feed intake significantly when the chopped leaves and pseudostems were supplemented with sweet-potato foliage or with wheat bran. The highest feedintake was observed when both sweet-potato foliage and wheat bran were given together.

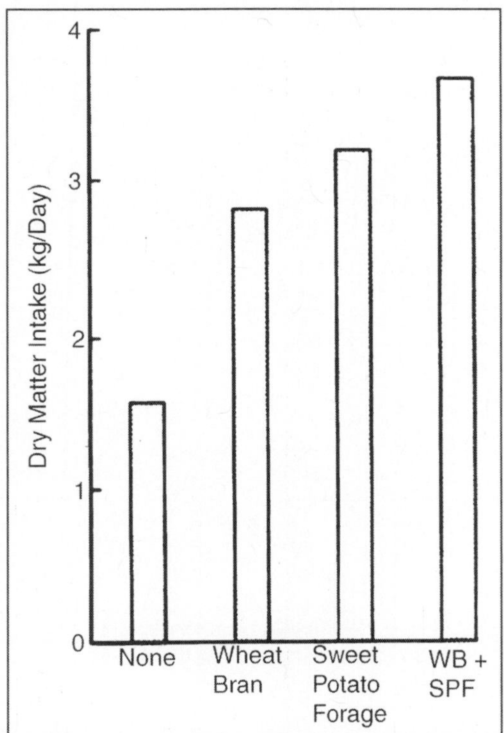

Fig. Supplements of Wheat Bran (1kg/d) and/or Sweet Potato Tops (5% Bodyweight, Fresh Basis) Increased the Voluntary Feed Intake of Cattle given a Basal Diet of Chopped Banana Pseudostcm, Urea and Minerals

Feeding systems based on banana foliage have been developed in the Seychelle Islands. Ac-ceptable levels of growth were obtained incattle fed the banana forage, urea and leucaena, and there was an apparent response to additional supplementation with banana fruit.

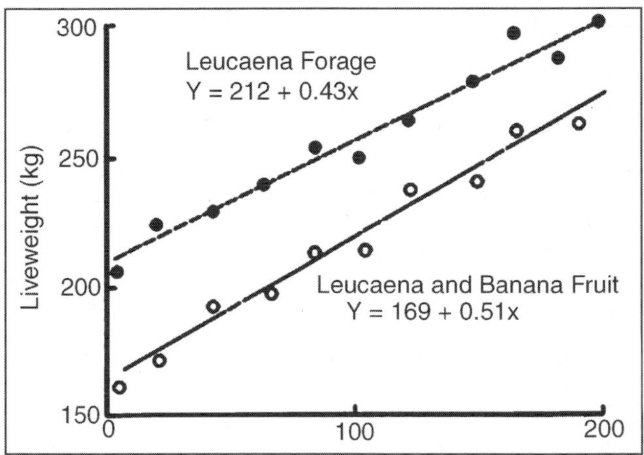

Fig. Growth Performance of Crossbred (Jerrsey/Creole) Cattle in Seychelles Fed a Basal Diet Based on the Whole Banana Plant (Pseudostem and Leaves Plus Urea and Minerals) and Supplements of Fresh Leuucaena Foliage (3% Bodyweight/d) Alone or with Some Reject Banana Fruit (3kg/d) (M Delpeche and T R Preston, unpublished data).

Sugarcane produces enormous amounts of biomass. Picture shows sugarcane being chopped for feeding trials with cattle in Mexico (Alvarez F)

Crop residues are the major feed biomass for cattle 111 developing countries.

SUGAR CANE

Growing sugarcane, more than any other crop, max-imises the yield of biomass per unit area.

Sugarcane is an ideal crop to optimise biomass utilisationbecause:

- Its C-4 pathway for photosynthesis confers both high-yield potential and efficient capture of solar energy

- The requirements of the large-scale sugar indus-try have promoted agronomic characteristics in the plant which facilitate fractionation
- Its perennial growth habit and disease resistance aid maintenance of soil fertility and reduce soil erosion, allowing monocultural practices with only minimal use of inputs derived from fossil fuels.

There are a number of sugarcane byproducts that can be used in animal feeds. Details of the two princi-pal methods of extracting sucrose fromsugarcane and the associated crop residues and resultant byprod-ucts are outlined in Figure. Of the two extrac-tion processes, the industrial technology produces the most byproducts, the principal one being molasses. In the artisan (small-scale) system for production of "Gur" (Indian continent) or "Panela" (Latin Amer-ica) there is no centrifugation and therefore no final molasses.

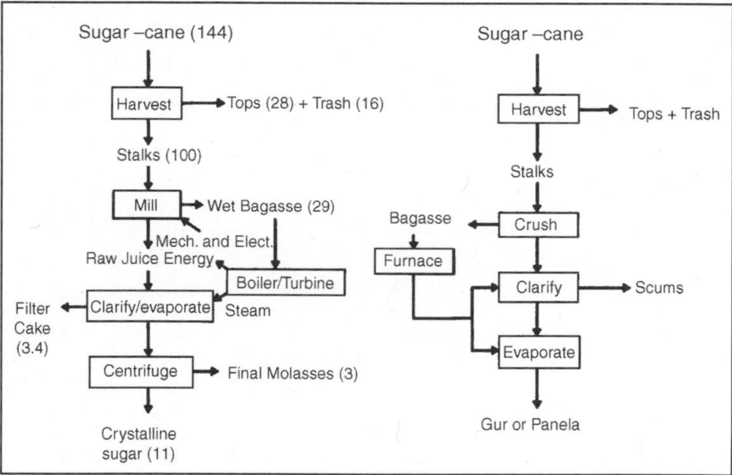

Fig. Residues and Byproducts from the Manufacture of Sugar by "Factor'y" and "Artisan" Methods. Numbers in Brackets Indicate Approximate Quantities (fresh basis) Relative to Cane Stalk = 100. I

WHOLE SUGAR CANE

Urea Supplementation

In all the trials discussed here, urea was added to the sugarcane at levels that would satisfy the needs of the rumen micro-organisms for fermentablenitro-gen (approximately 3% of the dry matter of the diet).

The optimum level was validated in an experiment in which urea concentrations were varied from zero to 4% of the sugarcane dry matter. All cattlewere supplemented with rice polishings (1 kg/day) and miner-als. All the parameters of animal per-formance increased curvilinearly withincreasing urea in the diet.

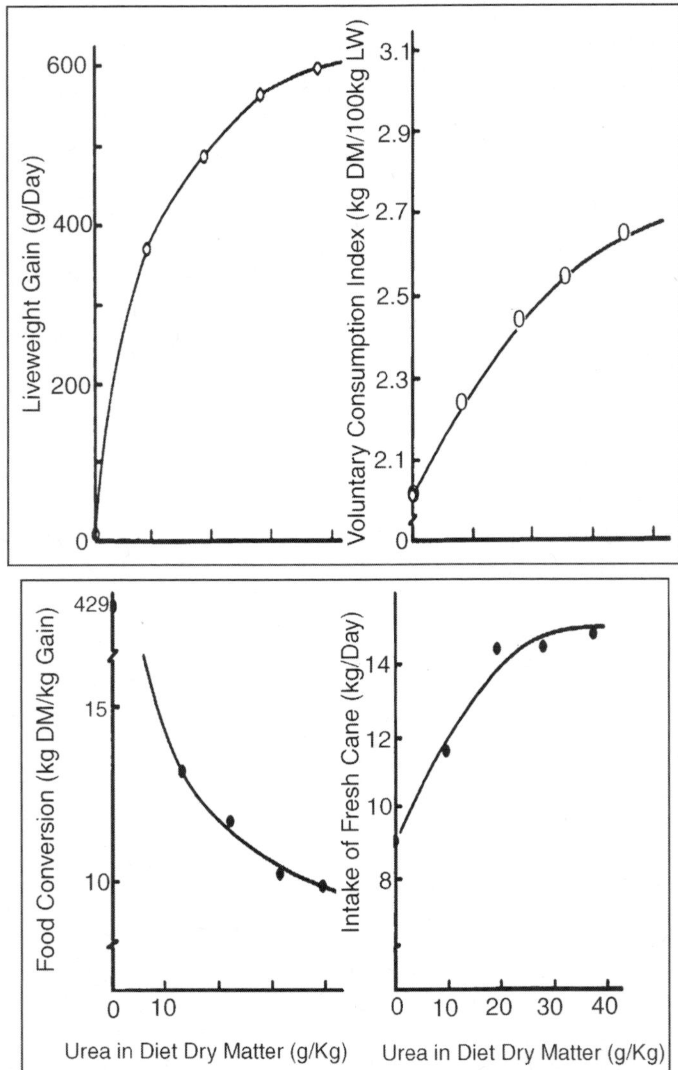

Fig. The Relationship between the Level of Urea in the Diet and Live Weight Gain, Feed Intake and Feed Conversion of Cattle a Diet of Sugar Cane and 1 kg/d of Rice Polishings (Alvarez and Preston 1976b)

When rice polishings were absent from the diet, sugarcane intake remained low and the animals lost weight irrespective of urea level; the only positivere-sponse was in digestibility (Ferreiro *et al. 1977).*

Cattle growth rates and feed conversions were sim-ilar when a urea solution was sprayed on to the cane (as an aqueous solution or in dilute molasses), or was given in a separate feeder as concentrated (10%) so-lution in molasses.

Foliage from food crops and legume trees

Cassava tops and leucaena forage given as supple-ments to cattle fed *ad libitum* chopped whole sugar-cane plus urea, increased the cattle's voluntary intake of the total diet but reduced the intake of sugarcane (Meyreles *et al.* 1977; Hulman and Preston 1981); improvements in growth rate were small(from -40 to +140g/day with cassava as the principal supplement; and from 60 to 200g/day when leucaena was fed).

A supplement of sweet-potato foliage added to a basal diet of chopped sugarcane and urea did not depress intake of the cane and increased total dry matter consumption by 34% (Meyreles and Preston 1978a). Liveweight gain was not measured in this trial but significant improvements in liveweightgain of cat tle were observed when this forage was added to a ration of derinded cane stalk (cane pith). This indicates that the foliageof sweet-potato is one of the most suitable "protein-rich" forages to use with sugarcane. The reason may be its higher rate of degradability in the rumencompared with ei-ther cassava or leucaena forage (Santana and Hovell 1979a, b), which results in a low rumen "load" while at the same time it stimulates rumen function.

Fresh banana leaves, which are only slowly digested in the rumen, were particularly unsuitable as a sup-plement as they reduced the intake of sugarcaneand resulted in a lower total dry-matter intake (Meyreles and Preston 1978b).A comprehensive series of experiments with sugar-cane as a basal feed for cattle was carried out in Mexico and the Dominican Republic with the aim of understanding the constraints associated with feeding this crop as the basis of the diet for growing/fattening and lactating cattle (Preston and Leng*1978a,b).*

The supplement that promoted the highest level of productivity was rice polishings given at a level of 10-15% of the diet dry matter. Similar conclusions were reported by Creek *et al.* (1976), who fed diets of 59% sugarcane (dry-matter basis) supplemented with ricepolishings and cottonseed meal to Boran (Zebu) cattle in Kenya.

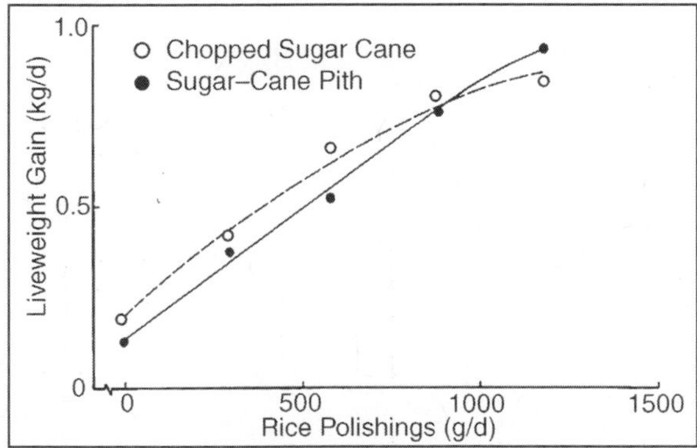

Fig. Effects of Supplementation with Rice Polishings (which Supplies Bypass Protein, Bypass Starch and oil) on Growth Tates of Zebu Bulls in Mexico Fat-tened on Basal Diets of Whole Sugarcane which has been Chopped or Derinded by the "Tilby" Separator Process

Rice polishings are relatively rich in protein(amino acids), lipid and starch, which are the critical nutri-ents that need to be adsorbed where the basal feed is digested solely by rumen fermentation. Experiments with cattle fitted with duodenal cannulae showed that the greater part of the starch (brokengrain) in the rice polishings bypassed rumen fermentation (Elliott *et al.* 1978a); the amounts of microbial and dietary non-ammonia nitrogen arriving atthe duodenum also increased in direct proportion to the amount of rice polishings in the diet (Elliott *et al.* 1978b) indicat-ing that rumen bacterial growthwas stimulated and also that protein in rice polishings was protected from rumen fermentation.

The development of a sugarcane-based feeding sys-tem for dual purpose (Brown Swiss x Zebu) cows and their calves was described by Alvarez and Preston (1976b) and Alvarez *et al.* (1977, 1978).

Best re-sults were obtained with combined supplementation of 500g of rice polishings per day andrestricted graz-ing (3 hours/day) on a leucaena "protein bank" (Al-varez *et al.* 1978). Complete substitution of leucaena for the rice polishings led tolower milk yields and loss of bodyweight in the cows which was attributed to mimosine toxicity (Alvarez and Preston 1976a) but may have been also aresult of a lower fat intake.

Sugarcane tops are the traditional feed for draught animals (cattle, buffaloes and mules) employed in the harvesting of cane and its transport to the sugar mil. The utilisation of cane tops by work-ing animals has apparently not been studied. Cattle appear to maintain bodyweight while carrying out quite arduous work on a diet of only sugarcane tops.

It appears that fermentative digestion of cane tops in the rumen provides an adequatebalance of nu-trients for maintaining bodyweight when the energy (acetate) available (relative to protein) is reduced by work. Implicit inthis statement is the concept that work increases feed intake.

Table. Composition of Sugarcane Tops and of the Rind and Pith Fractions Produced by the "Tilby" Pro-cess.

Pith	**Rind**		
Dry matter (DM) (per cent)	22	39	27
Composition (per cent DM)			
Protein (N×6.25)	1.4	3.2	2.7
Ether extract	0.2	1.0	0.8
Total sugars	46	24	27
Fibre	45	70	57
Ash	1.9	3.1	5.3
Sulphur	0.2	0.3	0.4

The apparent high nutritive value of cane tops for draught animals contrasts with their inability to sup-port growth in young animals withoutsupplementa-tion.

The data summarised in Figure show that cane tops support high growth rates of cattle only when appropriately supplemented.Growing Zebu steers given a basal diet of chopped sugarcane tops grew at almost 700g daily when this feed was supple-mented with urea and 1kg of rice polishings per day. Liveweight gain was the same on chopped cane tops as on chopped cane stalk. However, feed utilisation efficiency was higheron the cane stalk, apparently because of its higher content of solu ble sugars, which could have led to a larger proportion of propionate in the rumen VFAs.

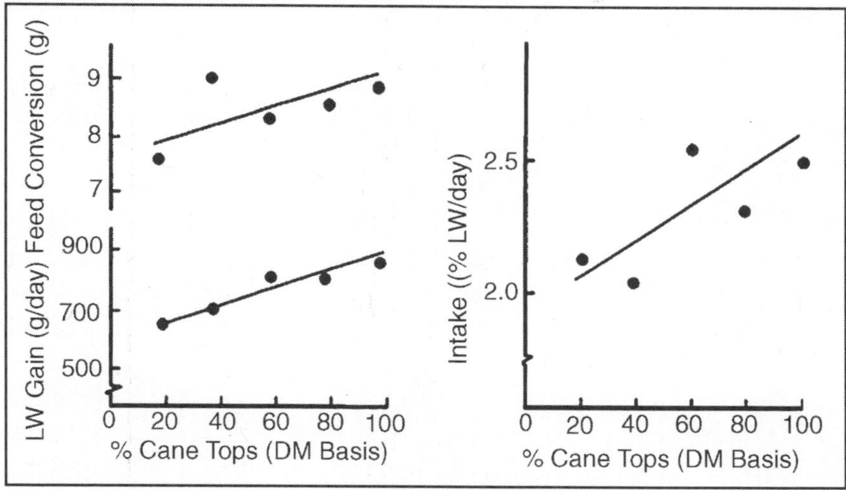

Fig. In a Fattening Diet f01' Cattle, Increasing the Proportion of Cane Tops (Replacingchopped Cane)

The constraints to the wider use of sugarcane tops in livestock feeding are the economics of harvesting and transporting the voluminous material.Tradi-tionally, sugarcane tops have been collected by small-holder farmers using draught animals, and even as "head loads". The increasing practice of burning the cane prior to mechanical, and even hand harvesting has reduced considerably the quantity of cane tops available.

Cane tops are one of the most under-utilised re-sources even in those countries in which shortages of animal feed and of fuel are most evident. Whenthey are fed to livestock, however, they have generally been used inefficiently because of the lack of knowl-edge of the need for "key" supplements. If they could be treated safely (and economically) to increase fibre digestibility this would improve their feeding value. Urea-ensiling has been effective in this respec.t (A. Boodo, unpublished).

In the late 1960s a technology (Tilby Separator Process-Figure) was developed for separating the cane rind from the pith. The aim was to use thecane rind in the manufacture of compressed boards which could substitute for plywoods that are usually imported into tropical countries. The residual pith was to be used for sugar extraction or alcohol production or as a feed for livestock.

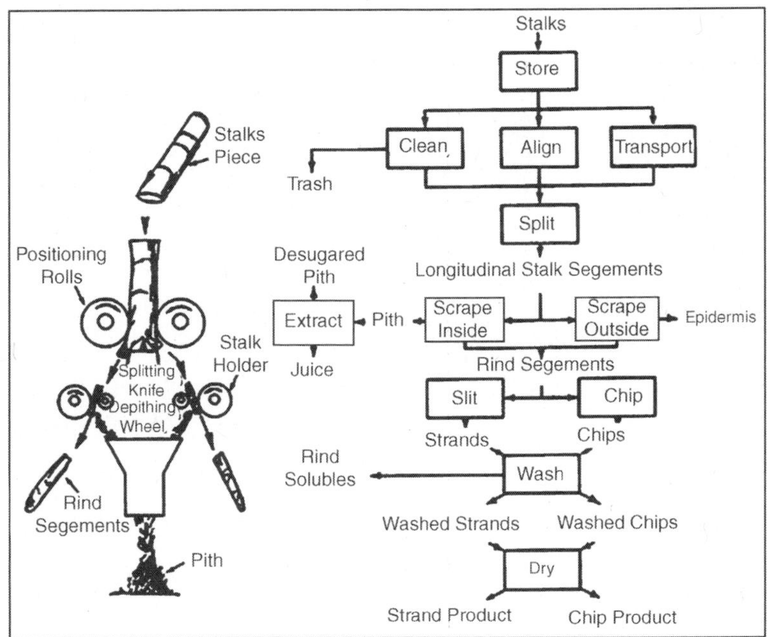

Fig. Simplified Diagram of the Tilby Separator Process for Derinding Sugarcane

In early experiments carried out by Donefer and his colleagues in Barbados, high rates of growth were observed in cattle fed sugarcane pith when it was supplemented with chopped cane tops and a concentrate containing an oilseed meal and urea. Unfortunately, thesepromising developments have not led to commercial application. The main limitation is the need for sophisticated and expen-sive machinery,necessitating a large-scale plant for economic viability. A factory producing compressed board would need a production capacity of about 30 tonnes ofboard per day, requiring the processing of 400 tonnes of sugarcane stalks. This would produce about 300 tonnes of pith per day-enough to feed15,000 head of cattle .. While this size of unit is not unusual by feedlot standards in the USA, it is quite impractical for most developing countries,where the use of sugarcane as an animal feed has most potential.

A second constraint which emerged from research in Mexico (Preston *et al.* 1976) was the realisation that, as a feed for cattle, the pith was little betterthan the chopped whole sugarcane plant when both were adequately supplemented, even though sugar-cane pith contains a higher concentration of soluble sugars than chopped whole cane and thus has a potentially greater feed value. In fact when it was supplemented with ricepolishings (providing bypass protein, starch and oil) at low levels, the whole sugarcane supported higher growth rates than the derinded stalk. Only at the highest level of supplementation was there an indication that ani-mal productivity was higher on a basal diet of pith. Nevertheless, differences were relatively small. These data indicate that it is the com-position of the supplement which determines animal productivity rather than minor variations in the ratio of sugar to fibre in the basal diet.Meyreles *et al.* (1979) demonstrated that there *is* potential for exploiting the higher energy value of the pith by rational supplementation.

In the absence of supplements other than urea and minerals, animals did not grow on a diet of sugar-cane pith. When green foliage of sweet potato was added to the diet, however, growth rate increased to about 500g/day and there was a significant re-sponse to high levels of urea (400 to 600g/ day).

This can be interpreted as indicating an increase in micro-bial growth in response to an improved rumen microbial ecosystem, which. may be due to thesupply of both readily digestible,fibre and essential nutrients for micro-organisms (eg.amino acids, peptides, min-erals and, perhaps, vitamins): Supplementation with cottonseed meal gave a similar growth response, but apparently without stimulating rumen function be-cause feeding the higher level of urea had no effect on growth rate.

The combination of these supple-ments was additive and feeding the high level of urea appeared to stimulaterumen microbial growth, as in-dicated by the 60% increase in growth rate.

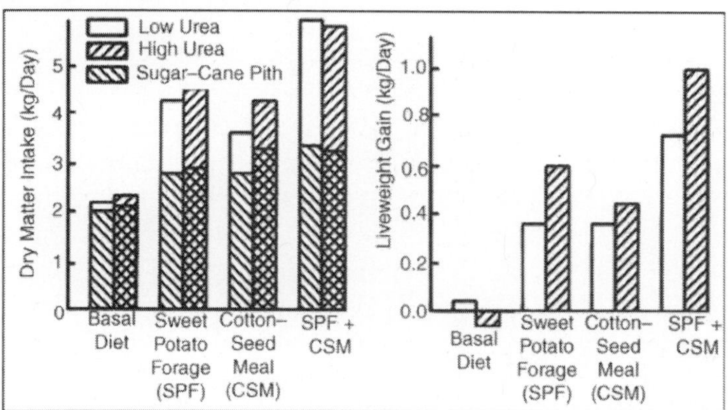

Fig. In a Cattle Fattening Diet Based on Ad Libitum Molasses, Sugarcane Tops and Wheat Bran (1 kg/d), Urea was a More Effective Source of Fer-mentable-N than Poultry Litter'. Poultry Litter Stimu-Lated Growth Rate when Added to the Diet (with Urea), which Wasappar'ently Adequate in Fermentable-N

Despite the possible advantages of using sugarcane pith rather than whole-cane, in some instances, the technology needed for chopping the wholesugarcane plant is simpler and much less sensitive to economies of scale than the derinding process. Chopping is therefore the preferred system for processing sugar-cane as a ruminant feed in developing countries.

FINAL MOLASSES

Alternative uses of Molasses as Animal Feed or for Power Ethanol

Molasses is the only "concentrated" source of fer-mentable carbohydrate that is widely available in the tropics and which is not a staple of the human diet. Its importance as a livestock feed is indicated by the ways in which it can be used:

- As a carrier for urea, minerals and other nutri-ents for improving the efficiency of utilisation of low-N diets (eg. crop residues, sugarcane and agro-industrial bypro ducts)
- As a strategic drought feed reserve and as the basis of a supplement for routine feeding during the dry season.

Molasses is traditionally used to manufacture potable alcohol (eg. rum). However, since the oil crisis enormous efforts have been made to developproduction of industrial alcohol, especially as a sub-stitute for gasoline (power-alcohol).

The lack of tech-nical knowledge of how to use molasses efficiently in livestock feeds has resulted in much molasses being discarded. Often molasses accumulates in the storage tanks at the sugar mills and is discarded into rivers. Its opportunity cost is thus very low, which has been a principal argument for its use in power-ethanol pro-duction.

However, the recent developments in the use of molasses in livestock feeding have given profitable alternatives that are vastly superior economically to the production of power alcohol.

Table. Molasses may have a Higher Economic Value when Used as an Animal Feed than when it is Converted into Power Alcohol

	Opportunity	Assumption	cost
$/t			
Survival feeding of cattle in drought	540	1	
Substitute for cereals	150	2	
Fattening cattle	92	3	
Supplementing crop residues/dry pastures for:			
Milk production	440	4	
Prevent body weightloss	47	5	
Power alcohol	15	6	

- To keep cattle alive for 6 months about 360~:g mo-lasses/urea (1%) is required. The cost of a replace-ment animal is estimated at US$200. After sub-stituting the urea and distribution costs the value attributable to molasses is approximately $540 per tonne
- The value of molasses as a replacement for imported grain i .• estimated on the basis that it has a substitu-tion value of 75 per cent and assuming that imported grain is US$200/tonne (CIF)
- It is estimated that used as the basal diet (70% of the total feed) for fattening cattle 6kg molasses supplemented with 150g urea and 8kg fresh legume forage supports a daily liveweight gain of 750g. The urea cost.. $0.045, the forage US$0.46 and the liveweight gain is valued at US$0.87/kg

- The daily con.oumption of 500g of a molasses/urea block; (50% molaHes, 20% urea, 20% bran, 10% minerals) ha.. increased milk yield on a straw-based diet by lkg/day (milk is worth to the farmers US$0.20/kg
- A daily intake of 1. 5kg of 8 per cent urea/molasses mixture prevents a weight 10H in cattle of 200g/day and that this weight 10H costs US$0.50/kg

4kg molasses are fermented to produce 1 litre of an-hydrous alcohol. Costs that are not included are the fuel for distillation and the interest/depreciation on the investment in a distillery. A barrel of oil (CIF) is assumed /,0 be worth $15.There is an extremely strong argument for retain-ing molasses as a livestock feed in countries that suffer dry seasons and periodic droughts. The opportunity cost of using molasses (which may be the only feed available) in drought regions in the tropics is related to the value of a live animal and thesurvival of subsis-tence farmers who are dependent on draught animals.

Molasses is used widely in compounded feeds in the industrialised countries, where it is used to improve the palatability and binding properties ofpelleted feed as well as reducing dustiness. Only low concen-trations (5-10%) are needed for this purpose; higher levels make mixing and pelleting difficult. Because the molasses comprises such a small proportion of the diet, no nutrient imbalance is apparent. In trop-ical countries the quantities of molasses available are large relative to other potential feed ingredients.

In these situations molasses can be used in three ways:

- As a fermentable carbohydrate providing the ba-sis of the diet for ruminants
- As a palatable carrier for other essential nutri-ents (eg. urea and minerals) to supplement fi-brous diets and also a gelling agent in the pro-duction of nutrient blocks
- As a source of trace minerals and some macro-elements (eg. sulphur, calcium and potassium).

Many attempts have been made to incorporate mo-lasses into relatively inert materials such as bagasse and bagacillo (the residue after extraction of the coarse fibres for paper and particle board manufac-ture) but the final product is of poorer feeding value than the original molasses and is more expensive due to the high cost of power for mixing.

Supplementation

Research was conducted in Cuba in the late six-ties aimed at developing livestock feeding systems in which molasses was the principal ingredient. Atthe outset it was decided that molasses should be fed in its original liquid state in order to reduce processing costs and to facilitate transport and storage. The successful development of the high-molasses fattening system for cattle

(Preston *et al.* 1967) ex-emplifies the use of the basic principles ofruminant digestion and metabolism as outlined in earlier chap-ters. These include optimising rumen fermentation by supplying fermentable N (urea) andsome high-quality green forage and balancing the amount of pro-tein (amino acids) to VFA by supplementing with a bypass protein source.The original system used forages such as ele-phant grass, pangola grass and sugarcane tops as the roughage source. The availability of forage was restricted to 0.8kg dry matter/100 kg liveweight to encourage the animal to consume large amounts of molasses. The urea level was set at 2.5% of thefresh weight of the molasses to provide a ratio of fermentable N to carbohydrate close to the theoreti-cal requirements of rumen micro-organisms.Sulphur supplementation was not required as sulphur diox-ide used in clarification of cane juice and the residual sulphur concentrated in the molasses.In the widespread commercial application of the feeding system in Cuba, fish meal (Peruvian) was the bypass-protein supplement. The dramatic effect of this supplement in raising animal productivity on the molasses-based diet is shown in Figure.

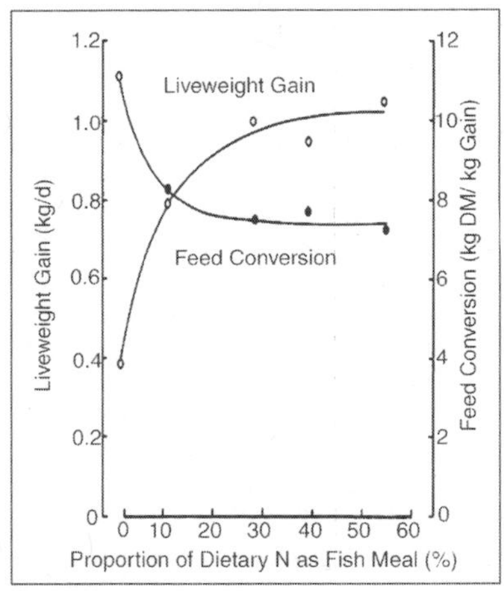

Fig. Addition of Fish Mealk to a Basal Diet of Ad Libitum Molasses-Urea and Restricted Forage Substantially Improved Growth Rate and Feed Conversion of Cattle in Cuba

Subsequent developments in the use of molasses-based diets have been directed to the use of:

- Protein-rich forages that supplied much, and some-times all, of the bypass protein as well as the roughage ; and
- Supplementation with poultry litter.

Table. Leucaena Foliage was as Effective as Groundnut Meal in Supporting High Growth Rates in Young Friesian Bulls Fed a Diet Based on Mo-lasses/urea. The Leucaena Replaced both the Forage (Native Grass) and the Groundnut Cake, and May thus have Served as a Combined Source of Protein and Roughage

	Leucaena (per cent LW/d)			**Groundnut.**	
Cake (g/d)					
LW gain (g/d)	790	740	847	597	742
Feed conversion	9	12	10	10	10(kg/kg gain)

Molasses (per cent of diet)	79	68	62	73	53

Table. The Foliages of Cassava or Sweet Potato was used as the l'oughage Supplement in Amo-lasses-urea Diet for Fattening Bulls in the Presence or Absence of a Supplement of Soyabeanmeal. With Cas-sava Supplementation there were no Benefits from Ad-ditional Protein Indicating that it was a Better Source of Bypass Protein than Sweet Potato Foliage.

Forage	**Sweet potato foliage**		**Cassava foliage**	
Soybean meal (g/d)	0	400	0	400
LW gain, g/d)	650	850	850	870
Molasses (kg/d)	5	6	6	6
Forage (kg/d)	13	13	10	10

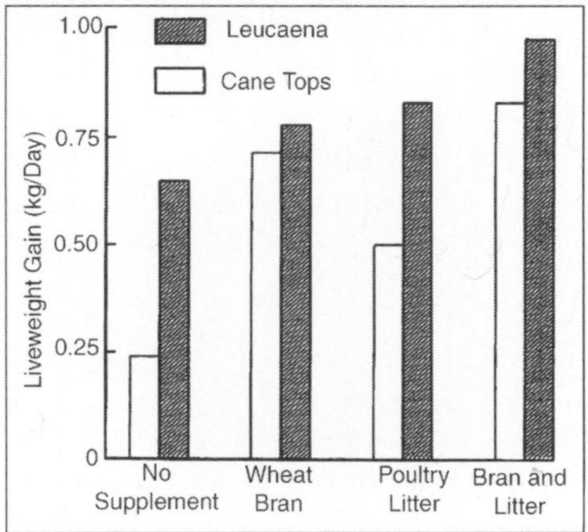

Fig. Contrasting Effects of Legume (Leucaena Foliage) and Grass (Sugarcane Tops) as the Roughage Supplement in a Diet Based on Ad Libitum Molasses/u1'ea with or with out Supplements of Wheat Bran

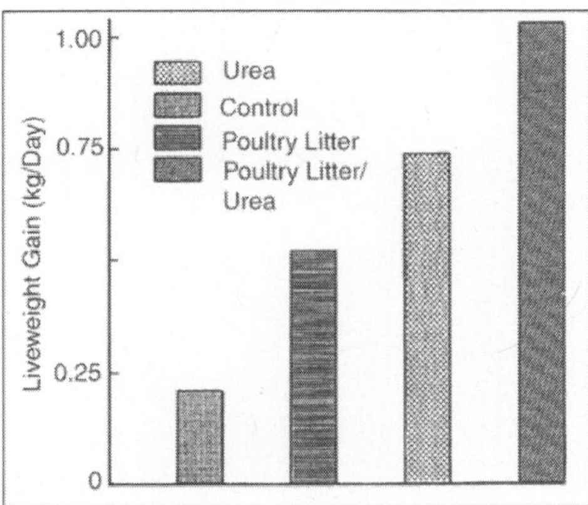

Fig. In a Cattle Fattening Diet Based on Ad Libitum Molasses, Sugarcane Tops and Wheat Bran (1 kg/d), Urea was a More Effective Source of Fer-mentable-N than Poultry Litter'.

Poultry litter stimu-lated growth rate when added to the diet (with Urea), which was appar'ently adequate in fermentable-N. There is evidence that, on molasses-based diets, poultry litter influences the pattern of VFA forma-tion, increasing the proportion of propionate and de-creasing the proportion of butyrate. This would partly explain the improved growth rates and feed conversions associated with the use of poul-try litter in molasses-based diets.

Commercial Feeding Systems Based on Molasses

The data m Tables summarise the results obtained when the molasses fattening pro-gramme was used commercially in large-scale feedlots and under conditions of restricted grazing on State farms in Cuba.

Table. Production Data from a Commercial Cattle Feedlot (10,000 Head) in Cuba During the Years when a Change was Made from an Ad Libitum Forage(Pan-gola Grass) Plus Concentrate Diet to one Based on Ad Libitummolasses/urea and Restricted Forage and Fish Meal. The High Losses Due Toemergency Slaughter and Mortality in the First Year of High Molasses Feeding were Due to Molasses Toxicity.

	Forage	Molasses-based	
	1969	1970	1971
Total LW gain (kg/d)	3,724	8,295	13,797
LW gain (g/head/d)	430	880	890
Feed conversion(kg DM/kg gain)	15	11	10
Deaths (per cent)	0.1	1.0	0.2
Emergency slaughter (per cent)	0.1	1.0	0.2

Table. Performance of Bulls Fattened Commer-cially in Cuba on Ad Libitummolasses/urea, Fish Meal and Restricted Grazing (3 hours/ day). Results are for 11 Unitseach with 400 Animals

	LW gain (kg/d)	Conversion (kg/kg LW gain)		
		Molasses	Fish meal	Urea
Worst	0.74	14.7	0.54	0.47
Mean	0.83	9.1	0.45	0.29
Best	1.04	5.9	0.32	0.19

METABOLIC DISORDERS ON MOLASSES-FEEDING SYSTEMS

Three metabolic disorders may occur in cattle and sheep fed diets containing more than 50% molasses:

- Urea toxicity
- Molasses toxicity
- Bloat.

Urea Toxicity

This is discussed since, when high levels of molasses are used, urea is almost invariably included in the mo-lasses, and urea intakes may be as high as300g/ day (eg. in a 500kg dairy cow consuming 10kg of the molasses/urea mixture per day).

However, there is little risk of urea toxicity since the sugars in molasses and ammonia from urea are quickly used in rumen microbial cell growth.Animals that have not previously consumed urea can be safely permitted free access to molasses containing up to 3% urea without fear of toxicity.Toxicity will only occur if the urea is not uniformly distributed, or mistakes have been made in formulation.

Molasses Toxicity

This is probably the most serious problem associated with molasses feeding. For example, in the first year following the introduction of the molasses/urea fattening system in Cuba, mortality and emergency slaughter rates in a 10,000 head feedlot increased from 0.1 per cent and 0.4% (when a forage-based diet was fed) to 1% and 3% respectively, when the diet was changed to high levels of molasses/urea.

Cattle suffering from molasses toxicity salivate, stand apart in a "dejected" posture, usually with their heads lowered; and frequently are found "leaning" against the fence or feed trough. Invariably, eye-esight is affected and often the animal is blind. When disturbed they have an unsteady and uncoordinated gait and this led the "cowboys" in Cuba to refer to the affected animal as "borracho" (ie. drunk!!).

The nervous symptoms and blindness that were a feature of molasses toxicity indicated damage to the brain and it was subsequently shown by Verdura and Zamora (1970) that the clinical syndrome was indistinguishable from that of cerebro-cortical necrosis (CCN) also known as polioencephalomalacia (Edwin et al. 1979). The necrosis in the brain is readily seen and this allows rapid diagnosis.

The cause of the necrosis is likely to be a decrease in the energy supply to the brain because of either an absolute deficiency of alimentary thiamine, binding of thiamine analogues produced in the rumen and/or through the action of thiaminase in the rumen (Edwin et al. 1979), or a deficiency of glucose (Losada and Preston 1973). The evidence for the last explanation is that rumen propionate levels were extremely low (less than 10% molar) when the disease was induced in cattle deprived of forage (Losada and Preston 1973), and that daily oral administration of 400g of glycerol (a glucogenic substance frequently used in the treatment of ketosis) prevented clinical symptoms of the disease (Gaytan et al. 1977). In contrast, administering thiamine either intra-ruminally or intra-amuscularly did not prevent the disease (Losada et al, 1971).

Recent work by Rowe et al (1979a) introduced another element into the aetiology of molasses toxicity. They confirmed that removing the forage from a molasses-based diet induced symptoms of molasses toxicity, including brain necrosis, in which neither blood glucose level nor the rate of entry of glucose into blood was impaired. The most significant finding was that the absence of forage in the diet caused a rapid reduction in the rate of turnover of rumen fluid to the point of almost complete rumen stasis (from 1.5 to 0.5 volumes/day). Rowe et al. (1979a) hypothesized that, under these conditions, the supply of both amino acids and B-vitamins to the animal would be drastically reduced with a consequent decrease in the availability of thiamine for brain metabolism. Another explanation is that low rates of turnover of ru-men fluid might. encourage the proliferation of slow--growing bacteria that produce thiaminase.

It is apparent that a number of factors interact in the development of molasses toxicity. The basic defect common to all situations isthe disruption of energy supply to the brain. This could be brought about by inadequacies in the ab-solute supply of blood glucose or thiamine (requiredfor brain metabolism of glucose) or both, as well as by the action of thiaminases present in the feed or produced by the rumen microorganisms.

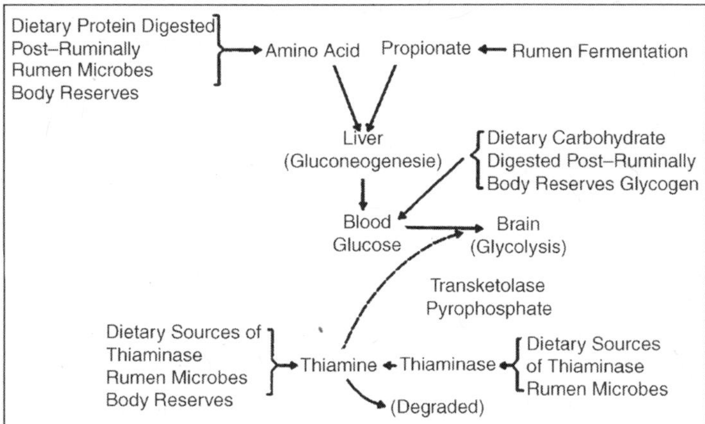

Fig. A Summary of some Factors that may Affect the Supply of Blood Glucose and Itsutilisation by the Brain

The theory that glucose deficiency is the main cause of molasses toxicity seems the most plausi-ble. The availability of glucose to cattle on molasses-based diets is always precariously balanced because of the low-propionate/high-butyrate fermentation of molasses, and t.he economic pressure to mini mise true-protein inputs. Molasses toxicity does not ap-parently occur in sheep. Sheep fed such diets have higher propionatefermentation patterns than c.attle which lends support to the concept of glucose being primarily involved in the syndrome. Removal of for-age from amolasses-based diet givell to sheep does not result in the same rumen stasis as in cattle.

Two processes may separately, or in combination, lead to excessive use of glucose and a lowering of blood glucose concentrations, which could lead to brain damage.

These are:

- Periods of exercise/excitement, which illcrease the utilisation of glucose, especially by muscle
- Consumption of an excessively large amount of molasses at one time, leading to a requirement for glucose to clear the increased amounts of ac-etate absorbed. Molasses ferments very rapidly in the rumen

and animals normally eat "little and often" throughout the day. Disturbance in this pattern, brought about by poor manage-ment or inadequate trough space and therefore "dominance" by some animals, could be the trig-gering mechanism for the over-eating.

This thesis is supported by observations of periods of low blood glucose levels in cattle during experi-ments which involved intensive blood sampling (R A Leng and M H Ferreiro, unpublished; also Gaytan *et al. 1977).*

When feeding high-molasses diets to cattle it is usual to restrict the supply of forage, either to in-crease intake of molasses or because of the greater cost of forage compared to molasses. Inadequacies in either the quantity or quality of forage appear to *be* the main causes of molasses toxi-city. Thus, the incidence of molasses toxicity was less when wheat or barley straw, rather than sorghum for-age or maize silage, were used as the forage sources in molasses-based diets. Furthermore, there have been no reports of toxicity when high-protein forages (eg. leucaena and cassava alld sweet-potato tops) have been used. Equally, feeding nutritious and palatable forage appears to be the best cure for affected animals (T R Preston, unpublished data) .

Recent developments in the molasses-feeding sys-tem have emphasised the technical and economic ad-vantages of feeding high-protein forages, especially

from leguminous plants such as leucaena, as a com-bined source of both roughage and bypass protein. Including such forages in the diet is also likely toof-fer the most cost-effective means of avoiding molasses toxicity.

Bloat

Bloat is the retention in the rumen of gas, either free or trapped in foam, and occurs in cattle on almost all feeding systems. It is more prevalent withsome diets (eg. grazed legumes that contain high levels of soluble foaming agents such as saponins and proteins). It is not a partic.ularly serious problem with molasses feeding. However, relatively high incidence of bloat (up to 5%) has been reported where access to the molasses is occasionally limited(eg. grazing cattle allowed access to molasses/urea only once or twice daily) and where cattle have eaten relatively large quantities of molasses/ureaover short periods.

A hypothesis for the aetiology of the syndrome is as follows. In animals fed molasses-based diets, an organism grows in the rumen which producesconsid-erable mucilaginous secretions and which can reach high population densities. When theanimals con-sume molasses quickly, the rate of fermentation in the rumen is high (producing considerable quantities of carbon dioxide and methane), thepH falls and the bicar bonate in the rumen fluid is converted to CO_2. The very rapid production of gas in the presence of the mucilagenous organism forms a foamy bloat. A similar train of events is reported to cause feedlot bloat on grain-based diets. An optimum level of rumen ammonia accelerates fermen-tation and therefore bloat may be controlled by using a source of NPN that releases ammonia more slowly than urea, eg:chicken litter.

SUGAR CANE JUICE

The extremely low fermentability of sugarcane fibre and the negative effect this had on vol-untary intake of the overall diet was the reason for developing methods for fractionating the crop so that sugar and fibre could be treated as separate entities.

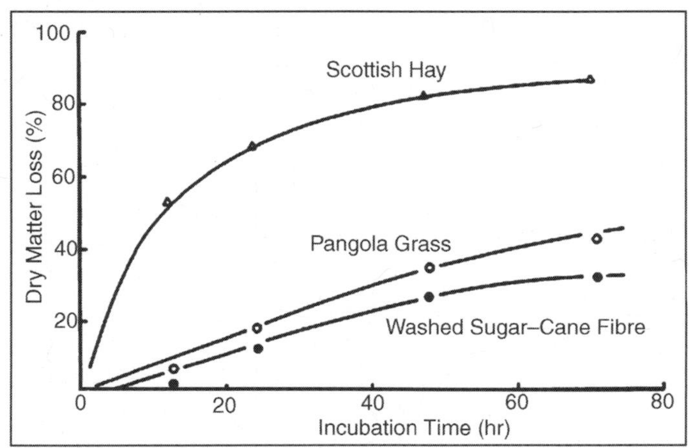

Fig. Rate of Loss of Dry Matter from Ny-lon Bags Containing Washed Sugarcane Fibl'e, Pangola Grass or a Sample of Hay Grown in Scotland, Suspendedin the Rumen of a Steer Fed a Sugarcane-based Diet

The new approach to fractionation of sugarcane for livestock feeding and fuel production was first devel-oped in Mexico (Preston 1980b). The aim was to achieve only partial extraction of the juice using sim-ple low-cost cane crushers, of the kind developed for "panda" and "gur" production. Thisreduced the investment in machinery and the energy cost of milling (stalks are passed only once through the crusher).

The justification for this system was that the juice is a suitable carbohydrate resource for both ruminant and non-ruminant livestock. The residual sugar-rich fibre can be gasified to make "producer gas" to sub-stitute for gasoline and diesel oil (T R Preston and A Lindgren, unpublished data). Alternativelyit can be used as a component of the diet of draught animals or small ruminants that can readily select the pith from the rind.

Supplementation

The fermentable carbohydrates in sugarcane juice and molasses are sucrose, glucose and fructose. Mo-lasses is the concentrated soluble residues afterex-traction of the sucrose from cane juice. It is there-fore richer in minerals, organic acids and other soluble plant components than cane juice.Diets based on cane juice support much higher lev-els of animal productivity than those based on mo-lasses. The rates of growth and feed conversioneffi-ciency recorded in studies in Mexico are comparable to those recorded in intensive grain-based systems.

Table. Holstein/Zebu Bulls had Higher· g1'Owth Rates and Better Feed Conversion on a Basal Diet of Fresh Sugarcane Juice than on Molasses. The Differ-ence was Especially Marked in the Absence of the By-pass Protein Supplement (Sunflower Seed Meal SSM).

Basal diet supplement(kg/d)	SSM (g/d)	LW gain	Intake of molasses/cane juice (kg/d)	Feedconversion (kgDM/ kg gain)
Molasses	0.0	250	4	26
	1.0	550	4	12
Cane juice	0.0	800	23	7.4
	1.0	1320	32	6.4

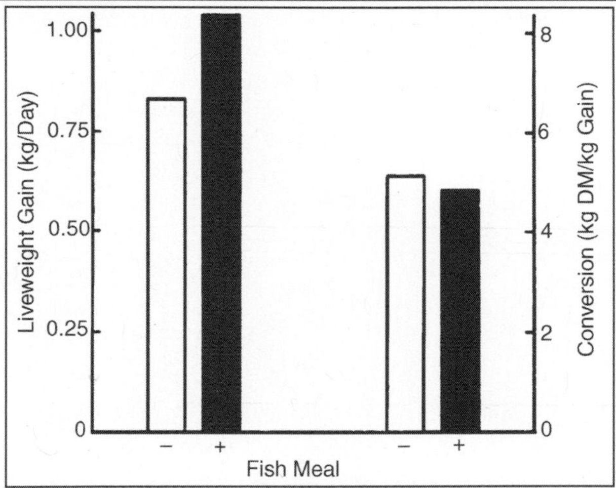

Fig. Zebu Steers had High Growth Rates and Efficient Feed Conversion Onsugar Cane Juice Supple-mented with Ammonia and Leucaena Foliage. Addi-tionalbypass Protein from Fish Meal Increased Growth Rate Slightly

The potential for use of cane juice is considerable because microbial growth in the rumen on this feed appears to be highly efficient; Zebu cattle grew at800g per day on minimal protein intake (5% protein in the diet dry matter supplied from leucaena) with the fermentable N provided by aqueousammonia. This is in contrast to the average grain-based feeding system, in which at least 12% of the diet dry matter is protein. It appears that on a canejuice diet, the end-products of rumen fermentation are better bal-anced and therefore support much higher levels of an-imal productivity than sugarcanepith. This indicates that microbial growth in the rumen was extremely efficient since it must have supplied an almost ideal protein/energy ratio for production without supple-mentation.

Two observations on feeding cane juice to cattle are pertinent to the underlying thesis of this book:

1. Propionate comprised more than 20% of the total VFAs in the rumen, giving a high glucogenic energy ratio (G/E) in the animals. In this respect the juice was dearly superior to molasses
2. The population of protozoa in the rumen of an-imals fed cane juice was lower than that. in the rumen of molasses-fed animals (ie. 10,000/ml versus 1,000,000/ml of rumen fluid), again suggesting efficient microbial protein production (D Harrison, personal communication; Bird and Leng1978).

The importance of these factors as determinants of efficient ruminant productivity are discussed.

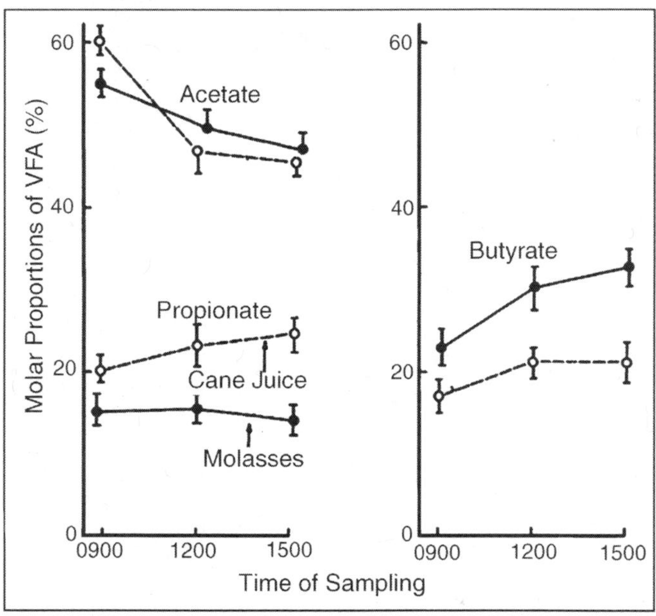

Fig. There were Higher Molar Proportion of Propionic Acid and Lowerproportion of Butyric Acid in Rumen Fluid of Cattle Fed Sugarcane Juice than with Cattle Fed Molasses

RESIDUAL PRESSED CANE STALK

The maximum economic benefit from the use of sug-arcane juice for livestoc.k feeding will only be re-alised iftechnologies are developed to use theresidual pressed stalk.

The options being pursued for the use of this material include:

- Productiojn of charcoal
- Conversion to producer gas (a mixture of carbon monoxide, hydrogen and nitrogen) as a substi-tute for diesel fuel
- Feeding to draught animals
- Feeding to small ruminants (sheep and goats), allowing a high degree of selection (eg. by pro-viding twice the normal dry-matter in take) .

Making charcoal from pressed cane stalk is simple but requires briquetting to produce a readily saleable product (Ffoulkes *et al. 1980).*

To be able to use pressed cane stalk for producer-gas production, it must first be sun dried to 20% moisture content and chipped into particles of 10 to 20mm in length. A gasifier has been designed that uses dried sugarcane chips (A Lindgren, personal communication).

A promising development is the potential of using the pressed (extracted) cane stalk as a basal diet for small ruminants. When it is available immediatelyfollowing juice extraction, goats eat it avidly, selecting the sugar-rich pith and rejecting the lignified rind. In studies in Colombia, goats on a mixed diet ofpressed stalk and fresh gliricidia foliage selected, and apparently preferred, the pith to the green foliage (Preston 1987).

CONCLUSION

In many developing countries, and in Asia in partic-ular, ruminants are fed on straw from cereal crops (mainly rice and wheat). As population pressure increases and the area devoted to food-crop produc-tion is extended the use of crop residues and byprod-ucts for animal feeding will increase. Incountries with specialised livestock production systems, straw is considered of such low feed value that it is often burned, but in developing countries where livestock are integrated with cropping it is a valuable resource.

Draught animals appear to be able to work and to maintain body condition on a diet of mainly straw. Research on the utilisation of straw by ruminants in developed countries has focused on the role of straw as a roughage supplement in a concentrate diet. This aspect of the use of straw is not important in the context of "matching livestock systems to available resources". The following discussion emphasises the use of straw as a basal diet and the use of supplements to increase productivity.

The nutritional value of straw varies according to a number of factors, including:

- Cereal species (in general, oat, barley, sorghum and millet straw are often of higher digestibility than wheat and rice straw)
- Variety and tannin content. There is a negative correlation between tannin content and digestibility (Saini et al 1977)
- Stage of harvest (eg. wetland rice is often har-vested green)
- Length of storage (which often depends on cli-mate and storage conditions)
- Proportion of leaf to stem selected
- Fertilizer application and soil fertility
- Irrigation: straw from irrigated wheat had an in vitro digestibility of41% compared with 34% for non-irrigated wheat (Kernan et al 1979)
- Plant diseases

- Weathering: leaf loss increases with time after removal of the grain
- Maturity.

Table. Straws from Various Cereal Crops Appear to be Highly Variable in Organic Matter Digestibility (in Vitro) (OMD). The N Content is Always Low

Wheat	28-58	0.4-1.0
Oat	34-68	0.4-1.0
Rice	40-52	0.5-1.0
Barley	34-61	0.4-1.0
Maize	31-50	0.5-1.2
Sorghum leaves	65-78	0.5-0.8

Table In vitro true organic matter digestibility (True-OMD) and lignin content of sorghum leaf and stem from 24 varieties grown under the same conditions. Among leaves, 80% of the variation in di-gestibility could be accounted for by the range in lignin content. Higher lignin content was associated with the presence of insoluble condensed tannins in the fi-bre fraction. Variation in digestibility of both leaf and stem fractions of sorghum straw is likely to influence productivity of livestock consuming it and should be taken into account by plant breeders.

True OMD(in vitro) (per cent)		Lignin (per cent DM)
Mean (range)	**Mean (range)**	
Leaf	70 (62-79)	6.0 (4.7-7.1)
Stem	65 (53-74)	6.6 (5.1-9.1)

All cereal straws have two characteristics in common:

1. Low nitrogen content
2. They are composed of cell wall components with little soluble cell contents and therefore have to be digested by microbial fermentation.

There appear to be no toxic substances in straws, except when they are mouldy. Rice straw, which forms a large proportion of the available forage in developing countries, contains high levels of oxalate and silica which are reported to be concentrated in the leaf rather than the stem. The lignin content of rice straw appears to be less than in other straws.

When straws are fed to ruminants the primary lim-itations to production are:

- The slow rate of, and low total digestibility
- The rate at which straw particles break down to a size that can leave the rumen
- The low-propionate fermentation pattern in the rumen
- The negligible content of both fermentable N and bypass protein.

12

Overlapping Eras of Cytogenetics Research

Progress of plant cytogenetics during the last ~80 years can be broadly divided in the following four eras, which may be slightly overlapping. Cytogenetics Era 1 (1910-1970): In this era chromosome number of a number of plant systems became known, structural changes like interachanges and inversions were studied for the first time in Stizolobium, Datura and Oentothera, inversions were studied in maize, and anuploids developed and cytogenetic maps constructed in several crops like maize, wheat, barley, etc. Alien addition and substitution lines were also developed in bread wheat using rye and few Aegilops/Agropyron species as the source of alien chromosomes.

Cytogenetics Era 2 (1950-1980): In this era, haploid DNA content (C-value) and composition (unique and repetitive) of nuclear DNA were determined in a large number of flowering plants using techniques of cytospectrophotometry and reassociation kinetics.

This led to recognition of two versions of C-value paradox. Firstly, the DNA contents in most eukaryotes were too high for the number of genes in the corresponding taxa, as estimated on the basis of known rates of mutations, and secondly the large-scale variation in DNA contents, could not be explained with the level of difference in complexity witnessed in these different organisms. The occurrence of large proportion of repetitive DNA in each of these eukaryotic genomes partly resolved the C-value paradox.

In this era, starting in early 1980s, DNA-based molecular markers were developed and molecular maps constructed in a large number of animals and plant systems, so that these molecular markers and the corresponding maps became an important resource for a variety of research problems, including their use in diagnostics and plant breeding. During this period, another significant development was the availability of a variety of fluorescence in situ hybridization (FISH), including multicolour FISH (McFISH), chromosome orientation FISH (CO-FISH), fibre-FISH, RNA-FISH, Comparative Genomic Hybridization (CGH) and 3-D FISH. This era started in mid-1990s and gained momentum in the present century, with two distinct areas of cytogenetics research; first, the whole genome sequencing giving birth to 'reverse genetics', and second, the chromatin remodeling giving birth to the concept of 'histone code'. It is thus obvious that chromosome research has evolved and progressed at an incredible pace in the last two decades, more particularly during the past ten years.

Since details of research done during Era 1, Era 2 and Era 3 would be widely available in text books and review articles, we mainly discuss the progress made during Era 4, in which significant progress has been made during the last few years to elucidate how the nucleosome and chromatin structure are modulated for expression of genes in time and space.

WHOLE GENOME SEQUENCING : GENOMICS AND EPIGENOMICS

In the year 2000, with the publication of the whole genome sequence of the crucifer weed, thale cress (Arabidopsis thaliana), plant cytogenetics entered into a new era of research, the era of plant genomics. Consequently in the early years of the present century, whole genome sequences of rice and a draft sequence of the genome of poplar (Populus trichocarpa) became available. Sequencing of several other plant genomes is also in progress (as listed at NCBI site) According to some, the first wave of plant genome sequencing is over, and we are now entering a new era in plant genomics research. In this new era, genomes of many model species with small genomes or those of species of economic importance will be sequenced.

Also the available sequences will be subjected to annotation (assigning functions to these sequences), and the choice of new genomes to be sequenced will be made on several criteria, including phylogeny.

The genomes of crops like maize and wheat will also be subjected to identification of gene-rich regions (GRRs), which will then be taken up for sequencing. There are also new technologies that will change the way we approach future genome sequencing projects.

DNA methylation, nucleosome remodeling (including histone modification and histone variants), and Non-coding RNAs can organize chromatin into accessible ('euchromatic') and inaccessible ('heterochromatic') sub-domains. This extends the information potential of the genetic code, and one genome can generate many 'epigenomes' in time and space, during the life-span of an organism.

The implications of epigenetic research are far reaching, so that efforts are being made to study the epigenomes in a variety of eukaryotes including some plant systems.

In a recent study, it was shown that these epigenetic modifications are not as conserved as was once thought. Further, very little is known about histone methylation in large genome plants, which make up the bulk of the angiosperms.

TRANSCRIPTION AND REPLICATION FACTORIES : A NEW CONCEPT

In recent years, specific regions have been identified within the eukaryotic nucleus, where replication and transcription takes place. It has also been shown that transcription sites are spatially distinct from replication sites. Therefore, we need to realize thattranscription by Pol II is not homogeneously distributed throughout the nucleoplasm, but occurs at highly enriched Pol II foci, known as transcription factories, which contain most of the hyperphosphorylated, elongating form of Pol II. Similarly replication takes place in 'replication factories'. It has been demonstrated that the chromosome segments destined to be transcribed or replicated have to move physically to these regions for transcription or replication to take place.

Transcription factories: A transcription factory is generally 80 nm tripartate structure mainly containing three spatially contiguous regions, with the template, the RNA polymerase II (Pol II) and the newly synthesised mRNA. The Pol II is perhaps attached to the transcription site, so that the template would move along RNAP II rather than the Pol II tracking along the template. Some translation also appears to be coupled with transcription at transcription factories.

There are fewer transcription factories than there are active genes and other transcription units in the nucleus, so that that more than one active gene is transcribed in each factory. And actively transcribed genes that are separated by long distances frequently co-localize in the same transcription factory. It has also been shown that actively transcribed genes co-localize with transcription factories, whereas identical, temporarily non-transcribed alleles, which can often be in the same cell, do not. Therefore, the 'on' state correlates with factory occupancy and the 'off' state with relocation away from factories. The specific Thus the specific nuclear repositioning of genes is correlated with transcriptional activation, silencing and replication timing.

The fact that different genes frequently co-occupy the same factory provides strong evidence that genes do not assemble their own transcription sites de novo when they become active, but instead migrate to preassembled transcription sites. A stable factory implies that genes or transcription units would essentially be pulled through a factory, rather than polymerases moving along the chromatin fibre, as is commonly believed. The finding that approximately 15% of the genome is transcribed — although probably not all at once — indicates that an extraordinarily large part of the genome passes through the limited number of transcription factories in a cell nucleus. This must have a profound effect on the nuclear organization of the genome.

Replication factories: It has been shown that a segment of DNA that needs to be replicated temporarily disengages Pol II and ceases transcription, although transcription in other regions of the genome continues uninterrupted throughout

S phase. This segment of DNA needs to relocate itself in a replication factory. There seems to be no overlap between replication and transcription sites/ factories.

CHROMATIN REMODELING

The term "chromatin-remodeling" generally refers to changes in histone-DNA interactions in nucleosomes. Histone proteins are modified and non-histone chromosomal proteins (*e.g.* high mobility group nuclear proteins = HMGN proteins; heterochromatin protein = HP1) are indirectly involved in the modification and activity of chromatin.

Histone-DNA interactions in the nucleosomes are also modulated (remodeled) to facilitate interaction of other factors (not involved in remodeling) with DNA template. It is believed that the histone octamers of nucleosomes are often displaced from the enhancer and promoter regions by chromatin remodeling complexes to allow access of a variety of factors to DNA. During 1990s, evidence for chromatin remodeling and cellular memory became available in yeast (Saccharomyces cerevisiae), fruitfly (Drosophila melanogaster), and mammals, through genetic and biochemical studies.However, later towards the end of the last century and in the early years of the present century, chromatin remodeling has been studied and factors associated with this phenomenon in plant systems like Arabidopsis.

Activities catalyzed by chromatin remodeling: Chromatin-remodeling factors can catalyze the following activities:

(i) Mobilization and repositioning of nucleosomes,
(ii) Transfer of a histone octamer from a nucleosome to a separate DNA template,
(iii) The facilitated access of nucleases (enzymes) to nucleosomal DNA,
(iv) Creation of di-nucleosome-like structures from mono-nucleosomes,
(v) Generation of superhelical torsion in DNA, (vi) disruption of histone-DNA contacts; and
(vi) Assembly and disassembly of nucleosomes.

Components of chromatin remodeling complexes: Three major strategies are used for chromatin remodeling. Each strategy makes use of a different set of proteins, described as components of chromatin remodeling complexes. Following are the three classes of chromosome remodeling complexes/ components:

(i) ATP-dependent chromatin remodeling complex, which makes use of ATP for chromatin modification through mere histone-DNA interactions.
(ii) Histone modifying enzymes, which are used for post-translational modifications of histones (mainly acetylation and methylation); these modifications create signals that define the so-called 'histone code'.

(iii) Variants of the histones (H2A, H2B, H3, H4 and H), which are synthesized and incorporated into nucleosome subunits (in place of normal histones), bringing about chromatin remodeling. More recently, the extreme use of H3 histone variants also led to the formulation of 'H3 barcode hypothesis'.

ATP-dependent chromatin remodeling machines/complexes: ATP-dependent :chromatin remodeling complexes (CRCs); also called chromosome remodeling machines (CRM), consist of ATPase polypeptide subunits and also non-ATPase subunits. Although ATPases play an major important role in chromatin remodeling, non-ATPase components also have a role to play. The ATPases can be quite diverse and not all ATPases are important for chromosome remodeling. These ATPases are classified into three superfamilies, each superfamily having several families. One of these families belongs to a specific class of ATPases, described as 'Snf2-like family of ATPases', which are relevant to chromosome remodeling and therefore will be discussed in some detail.

(a) ATPases as subunits of ATP- dependent complexes: Members of SNF2-like family of ATPases are classified in several subfamilies, depending upon which protein motifs outside the ATPase region they have. Atleast seven subfamilies are known, but only four subfamilies are important, which include the following:
   (i) Swi/Snf2 subunit of the SWI/SNF complex;
   (ii) ISWI (ISWI, hSNF2H, hSNF2L, yISW1, yISW2; h stands for human, and y stands for yeast);
   (iii)CHD1 (CHD1, Mi-2a/CHD3, Mi-2b/CJD4, Hrp1, Hrp3),
   (iv)INO80 (CSB, Rad 26, ERCC6).

(b) Non-ATPase subunits of ATP- dependent complexes: The ATP dependent chromatin remodeling complexes also have non-ATPase subunits, which may have the following functions:
   (i) enhance or regulate motor activity of ATPase subunits;
   (ii) mediate other specialized functions that are not related with chromatin remodeling– these may be recruited to promoters via interactions with sequence specific transcription factors (TFs). Rad54 is also an example, in which SNF2-like ATPase is programmed by another polypeptide. Rad54 and Rad51 (related with bacterial RecA protein) catalyze homologous strand pairing.

Chromosome remodeling through histone modifications: The 'histone code': In eukaryotes the fundamental unit of chromatin is the nucleosome, which is a protein octamer/DNA complex composed of 200 bp wrapped around a histone octamer that consists of two molecules each of four core histones,

H2A, H2B, H3 and H4. One molecule of linker histone H1 is also associated with each nucleosome. Crystal structure of the core particle of this nucleosome unit consisting of 146 bp of DNA wrapped around histone octamer was determined in late 1990s at 2.8 A resolution. More recently, crystal structure of a tetranucleosome was also determined at 9 Å resolution to understand the manner of higher-order folding of the nucleosome sub-units.

It was also shown in several recent studies that amino-terminal tails of histone proteins are targets for a series of post-translational modifications (PTMs), including acetylation, phosphorylation, and methylation. These modifications regulate chromatin structure and gene expression. Multiple histone modifications in various combinations are thought to form a 'histone code', since these modifications extend the information capacity of the associated DNA. For example, histone H3 and H4 acetylation is consistently associated with transcriptionally active euchromatin, while methylation can be associated with either active or inactive chromatin depending on the residue involved in methylation.

For instance, methylation at H3K4, H3K36 and H3K79 are hallmarks for active transcription, whereas methylation at H3K9, H3K27 and H4K20 are correlated with transcriptionally inert heterochromatin. Further, the lysines in histone N-terminal tails can be mono-methylated (me1), di-methylated (me2) or tri-methylated (me3), and each methylation state may have unique biological functions, increasing the potential complexity of the histone code.

Table. Histone modification combinations and their effect on transcription

Nature of histone Modification	Effect on transcription
Histone acetylation	Transcription activation
Histone methylation	
H3K4, H3K36 and H3K79	Transcription activation
H3K9, H3K27 and H4K20	Transcription silencing

Table. Various combinations of histone methylations and their effect on state of chromatin condensation

Degree of methylation	State of chromatin
H3K9me3, H3K27me1 and H4K20me3	Highly condensed heterochromatin
H3K9me1, H3K9me2 and H3K27me3	Lightly condensed heterochromatin

It has been shown that methylation of H3K9, H3K27 and H4K20 is involved in heterochromatinization, but the degree of methylation of each of these lysine residues determines the degree of heterochromatinization.

For instance, H3K9me3, H3K27me1 and H4K20me3 mark the most deeply stained regions while H3K9me1, H3K9me2 and H3K27me3 mark the less condensed heterochromatin. Several reports indicate that the monoand di-methylated forms of H3K9 and H3K27 are enriched in heterochromatin, although degree of methylation also depends on the genome size. For instance, unlike in animals, in Arabidopsis, H3K27me3 is associated with euchromatin and H3K9me3 is extremely rare. There is also evidence that H4K20 methylation is also present in Arabidopsis. It is not clear, however, whether there is a direct relationship between heterochromatin and histone methylation. However, in plants with large genomes including maize and barley and wheat, very little is known about histone methylation, which makes up the bulk of the angiosperms. In a recent study, quantitative distribution of mono-, di-, and trimethylation at H3K9 and H3K27 was examined in maize on whole genome level using oachytene chromosomes. The data reveal that three marks (H3K9me1, H3K27me1, and H3K27me2) correlate with DAPI (DNA) staining, but that only H3K27me2 is specifically enriched in condensed areas. It was also observed that H3K9me2 is not abundant in heterochromatin, but is instead enriched between chromomeres along with H3K4me2. The study also reported that centromeres contain H3K9me2 and H3K9me3; that H3K27me3 occurs at several brightly focused euchromatic domains, and that H4K20 methylation is rare or absent.

The Histone Code Hypothesis states that chromatin-DNA interactions are guided by combinations of histone modifications. For example, phosphorylation of serine residues 10 and 28 on H3 is a marker for chromosomal condensation; similarly, phosphorylation of serine residue 10 and acetylation of residue 14 on H3 is a tell-tale sign of transcription.

Chromosome remodeling through histone variants: 'H3 barcode hypothesis': In addition to post-transcriptional modifications (PTM) of histones, a number of variants are known for histone proteins that are found in the core particle of the nucleosome. Perhaps H4 is the only exception, which does not seem to have any variant, but H3 and H2A are the two major classes of histones, which exhibit higher variation relative to H2B histone. Each histone variants differs from its corresponding normal histone in only a few amino acid residues, and each histone variant plays an important role in chromatin remodeling.

In particular, the histone variants for H3 offer the best example to illustrate the role of histone variants in chromatin remodeling, so that a H3 barcode hypothesis has also been formulated recently to describe the role of histone H3 variants.

The histone variants are generally synthesized directly from genes encoding them, but may also result due to post-transcriptional modification of conventional histones. The genes encoding histone variants can be broadly classified into replication dependent (RI), replication independent (RI) and tissue-specific (TS), so that at least some of these variants occur in a tissue specific or developmental stage-specific manner. These histone variants replace the normal histones, and can be incorporated into the chromatin at any time during the cell cycle, although the conventional nucleosomes are produced and assembled into nucleosomes only during the S-phase of the cell cycle.

NUCLEAR ARCHITECTURE AND CHROMOSOMES

Chromosome organization within the nucleus: In the past, it was believed that chromatin within the nucleus is a network, intertwined and randomly distributed within the space available in the nucleus. However, both in plants and animal systems, recent evidence has demonstrated that the nucleus is a highly compartmentalized structure. Chromosome territories and interchromatin compartment (CT-IC model): The chromatin within the nucleus is organized in the form of chromosome territories(CTs) and interchromatin compartments (IC), and hence the formulation of CT-IC model. While CTs contain individual chromosomes, IC contains macro molecular complexes that are needed for replication, transcription, splicing and repair.

Newer techniques combining 3-D-FISH and computer aided deconvolution techniques helped in resolving the following features of chromatin organization and behaviour

(i) in an interphase nucleus, each individual chromosome occupies a discrete space, called the 'chromosome territory' or CT and that there is little inter twining among chromosomes;
(ii) in interphase cells, each chromosome also interacts with the nuclear envelope through consistent contact points;
(iii) in interphase cells, each chromosome interacts with other chromosomes through its heterochromatic regions; and
(iv) in dividing cells, chromosome movements are also non-random. The above information regarding chromosome organization at the physical level has also been integrated with genetic and molecular data, to decipher the mechanisms of different nuclear processes, in which chromosomes are involved, more particularly the transcription and DNA replication.

Rabl organization and telomeres orientation: Chromosome segregation at anaphase results in the polarization of chromosomes because sister centromeres are pulled in opposite directions and the rest of the chromosome arms trail behind. In some instances, this anaphase arrangement of chromosomes persists into the following interphase; this is known as the Rabl organization, in which chromosomes have a preferentially polarized organization, with centromeres at one end of the nuclear envelope, called the apical side, and telomeres at the opposite end, called the basal side. The presence of the Rabl organization is known to vary greatly between species and among tissues or developmental stages of an organism. In plants, it is generally observed in species with bigger genomes like those of wheat, rye, barley, and oats, but not in species with smaller genomes like those of sorghum and rice. Maize with intermediate size of genome displayed neither entirely Rabl nor entirely random chromosome organization.

Bouquet and telomeres clustering: The bouquet is the clustering of chromosome ends on the nuclear envelope (NE) during meiotic prophase, coincident with the initiation of homologous chromosome synopsis. The bouquet has been extensively described in many species in all eukaryotic groups and has been proposed as an aid to presynaptic alignment of homologous chromosomes. The similarity of the bouquet to the Rabl conformation has long been noted. However, it is clear that the two are not the same, although similar functions (chromosome pairing, recombination initiation and SC formation) have been assigned to both of them.

Relation between Chromatin Structure/ organization and Transcription

Heterochromatin and euchromatin: We know that the chromosomes of eukaryotes consist of darkly stained heterochromatin and lightly stained euchromatin. According to the classical view, heterochromatin is transcriptionally inactive and euchromatin is active.

This view is changing now; following are some examples: (i) there is evidence that several repeats within heterochromatic centromeres are transcribed to produce siRNA; (ii) it has been show that in each of the five chromosomes of Arabidopsis, heterochromatin is largely confined to the pericentromeric regions, and mainly consists of 180-bp satellite repeats, and retrotransposons (mainly from Athila family);(iii) NORs, mainly consisting of repetitive DNA, carries thousands of kilobases of tandemly repeated ribosomal DNA encoding rRNA. It has now been conclusively proved that in eukaryotic chromatin, cytosine methylation in repetitive DNA and distinctive modifications of individual histone proteins are frequent and cause heterochromatin formation.

There are at least three kinds of methylases causing cytosine methylation in DNA: (i) Dnmt-1 type DNA methylase adds methyl-residues to cytosines (CG or CNG?) on newly synthesized strand of a DNA duplex, by using information from the conserved parental DNA strand carrying met-C, thus making DNA methylation heritable; (ii) Dnmt-3 type DNA methylase brings about DNA methylation (CG or CNG?) de novo, so that unmethylated DNA becomes methylated on both strands; (iii) chromomethylases (unique to plants) in Arabidopsis and maize recognize CpNpG sequences and bring about cytosine methylation.

MOLECULAR CYTOGENETICS

Major genome research programmes are underway to isolate expressed genes of corn and to DNA sequence at least the gene-rich regions. We are developing a highly efficient method for mapping DNA sequences to chromosome and sub-chromosomal regions based on lines derived from crosses between oats and corn. Partial hybrids have been obtained that contain an entire set of oat chromosomes along with one corn chromosome; lines are available that have each chromosome of corn individually added to the oat genome.

These lines are now being used to generate derivatives with only a subregion of the corn chromosome. This genetic material as well as B73xMo17 derived lines will be used to learn more about the subgenomic structure of corn and to test whether the degree of heteroallelism within and between the subgenomes contribute to heterosis. A pathogenic bacteria, called E. coli O157:H7, can contaminate hamburger and cause serious hemoraging. The contamination largely comes from the feces of cattle. Strains of E. coli have been selected that will kill the O157:H7 strain.

Two genes appear to be involved and this project will test their efficacy in corn. Wild rice is a crop under domestication. The seed falls (shatters) from the panicle upon maturation and is not available to be harvested. Storms can cause major losses of wild rice yield at harvest.

The shattering trait is considered a domestication trait that has been modified years ago in most major crops; the trait still exists in wild rice varieties. We have discovered major genes for shattering and have molecular genetic markers that should allow more efficient breeding for shattering resistant varieties. Methionine is an essential amino acid required in the diets of non-ruminant animals. Poultry have high demands for methionine accounting for why synthetic methionine is added to poultry rations in the U.S. A line of corn was selected that is high in methionine. Several generations of crossing, backcrossing, and selfing has resulted in relatively homozygous lines potentially high in methionine,

The specific objectives are:

1. Develop efficient mapping system for corn DNA sequences.
2. Learn more about the subgenomic structure of corn and relate the information to productivity.
3. Manipulate corn to produce compounds toxic to an enterohemoraghic bacterium (E. coli O157:H7).
4. Provide molecular genetic methods to the breeding programme to enhance the selection of varieties of wild rice with improved seed shattering resistance.
5. Finalize the selection and release of high methionine lines of corn.

We have produced lines of oats with a single corn chromosome added. These addition lines are available for all 10 corn chromosomes, individually present in the oat background. These lines are being used extensively for mapping DNA sequences to chromosome.

The oat-corn addition lines are being irradiated to produce radiation hybrid (RH) lines that have only part of the corn chromosome present. These lines allow much more refined mapping of corn DNA sequences. Future research will involve continued development and characterization of the RHs, mapping ESTs, cytogenetic transmission studies of the various lines, and obtaining the transmission of the chromosome 10 addition and formation of RHs.

Molecular genetic analysis of corn has shown that its genome is composed of major duplicated regions suggesting an ancient allopolyploid origin. The potential exists for many loci to not only be duplicated but to have different alleles at each locus. Maximum productivity could be the result of maximizing the number of alleles at these loci. We will screen several genetic markers in order to determine Mo17xB73 lines containing duplicate regions containing various combinations of the parental genotypes.

Yield tests will ultimately be performed on selected lines with maximum heteroallelism in regions known to possess major yield QTLs. Our molecular genetics research on wild rice has resulted in the first molecular marker map and the placement of QTLs for several traits on the map. A PCR based marker has been developed that tags the most major QTL with polymorphism between shattering and non-shattering types.

This QTL accounts for about 40% of the variation in seed shattering in wild rice. A marker-assisted selection programme will be initiated using the molecular genetic marker for the major QTL and others as possible. Escherichia coli O157:H7 is a pathogenic strain of E. coli found in the digestive tract of beef cattle and causes food borne illness through contamination of beef during the slaughtering process. The goal of this project is to target the pathogen at the feed source or farm level to prevent contamination by developing a transgenic maize line that expresses a protein (termed "colicin") toxic to this strain of E. coli.

Colicin (produced by the cea gene) degrades DNA, and an immunity protein (produced by the cei gene) is required to prevent non-specific DNA degradation. Thus both genes are required to accomplish our goal. A screening procedure was developed in our lab that provides a screen for corn with elevated levels of methionine in the endosperm. A line was discovered, called BSSS5, that is resistant to lysine-threonine inhibition as a whole kernel but is inhibited when only the embryo is cultured.

The endosperm in this case possesses sufficient methionine to prevent the inhibition. BSSS 53 was found to produce a zein-2 protein with 23% of the amino acid residues as methionine. We have attempted to backcross the high-methionine trait into three different genetic backgrounds (B73, Mo17, and A632) and have some lines almost ready to release. Oat-corn addition lines have been developed for every corn chromosome. These strains allow - in a single experiment - the rapid mapping of corn DNA sequences to their respective chromosome.

These strains resulted from more than 100,000 oat x corn crosses and isolation of over 7300 immature embryos. We extended the available genetic backgrounds by recovering oat-corn addition lines with B73 or Mo17. Between the two inbreds, we have a complete addition line set; the B73 addition lines will be useful in assembling the corn genome.

The DOE Joint Genome Institute will sequence chromosome10 using our Mo17 chromosome 10 oat-maize addition line. We also have added to our collection of radiation hybrid lines for each chromosome. These lines allow the placement of corn genes to specific regions of the chromosome. Corn gene expression/silencing in the oat-corn addition lines is being tested using the DNA chip technology. Out of 17,000 tested sequences, 890 were expressed in the chromosome 5 addition.

About half of these sequences are of known position and are either placed on chromosome 5 or in a duplicated region of the genome. We further pursued the possibility of introducing C4 photosynthesis to oat with these materials.

Addition lines with both chromosomes 6 and 9 (carrying genes for key C4 enzymes) were tested for an effect on CO2 compensation point. A statistically significant change occurred in compensation point but not an impressive one. We will next investigate a third chromosome that has an effect on leaf cell morphology (known as Kranz Anatomy) involved in photosynthetic efficiency.

NIR analyses of self-pollinations of a new high oil accession indicate that the kernels contain 20% oil. A QTL analysis is underway. There is considerable interest in this material as a source of biodiesel and ethanol.

Using the molecular genetic map we published for wild rice, a PCR-based marker system has been identified for use in marker assisted breeding for the seed shattering trait. Genetic tests indicate a close marker/trait linkage. The recent outbreak of the pathogenic E. coli 0157:H7 on spinach emphasizes the importance of having control measures to minimize such occurrences. We are developing what might be considered a prototype approach of using corn feed to alter the microflora in the intestinal tract as a means to reduce the presence of this pathogenic bacterium at its source.

Whole corn plants were regenerated from appropriate calli and found to carry the colicin gene, express the RNA and protein, and have anti-microbial activity against E. coli O157:H7. An extensive bioinformatics analysis has been performed of 1100 DNA sequences involved in human cancers to determine if they are present in plants. P53 is a protein involved in more than 50% of human cancers. We found a p53 binding sequence in the promoter of the MAD1 (Mitosis Arrest Deficiency) gene in human; MAD1 acts as a mitotic checkpoint to ensure chromosome stability. A MAD1 homolog was found in maize. Laboratory analyses of knockout mutants of MAD1 in Arabidopsis are underway.

Crop productivity is at least half due to genetic advances. Continued understanding of genetics and the application of molecular genetics will allow continued progress in food production. In addition, additional health benefits and identification of unique biofuel resources will be forthcoming.

POLYTENE CHROMOSOMES ALLOW BANDING AT A VERY FINE LEVEL OF RESOLUTION

In specialized cells of some plant and animal species, chromosomes undergo many cycles of DNA replication without undergoing mitosis. The result is polytene chromosomes, which may contain as many as a thousand parallel copies of each chromosome, perfectly aligned along their entire length. Polytene chromosomes can be stained and viewed in interphase nuclei. The most common example is the polytene chromosomes in the salivary glands of *Drosophila.* Because interphase chromosomes are more extended than those in mitotic cells, loci can be identified on polytene chromosomes at a much finer level of resolution.

FLUORESCENCE IN-SITU HYBRIDIZATION

Fluorescence In-situ hybridization (FISH) uses fluorescence to highlight the site of specific DNA sequences. Trask *et al.* 1993 provide a detailed review of the technique and its applications. In essence, a DNA sequence is 'tagged' by incorporating fluorescently-labeled nucleotides into the DNA chain. The tagged DNA, referred to as a probe, is incubated with chromosomes on a slide. Wherever the chromosome contains the same sequence as the probe, the probe can base-pair with the chromosome. The site to which the probe base pairs will fluoresce in UV microscopy.

In-situ hybridization is a valuable technique for studying chromosomal distribution of specific nucleic acid sequences and to determine the location of highly repetitive DNA sequences. Changes in these highly repetitive sequences reflect the physical rearrangements in the genome over time in the evolution of the species. Highly repetitive DNA sequences represent between 20 and 90% of the DNA in higher plants. The rate of evolution of individual repetitive sequences varies considerably; some may be conserved and be present in many species within a taxonomic family. The original technique published by Gall and Pardue (1969) has undergone numerous modifications and fluorescent in-situ hybridization (FISH) has been applied to chromosome identification and diagnosis in human genetic studies.

As with Geimsa banding, chromosomes must be fixed before denaturation. It is necessary to denature both the DNA probe and the target chromosome to allow complementary pairing to take place.

Fixative: The chromosomes are fixed in a 3:1 methanol:acetic acid solution. Slides are prepared by the squash method and the cover glass is removed to allow the cells to dry. The squash method stabilizes the chromosomes, permits good penetration of the reagents and brings most of the chromosome into the same focal plane, and also increases the speed at which the hybridization takes place, by decreasing volume.

Denaturation : incubate at 72° C in 50-70% formamide and 2X SSC (NaCl) (formamide denatures DNA helices)

Addition of Probes: The probe molecules are pieces of DNA, selected to target specific sequences on the chromosome. The target sequences can be as small as 1kbp or larger than 15 kbp. Sequences can be taken from entire chromosomes to highlight large regions of the genome. Denatured probe is hybridized with fixed chromosomes under a coverslip overnight at 37° C in 50% formamide, 2X SSC and 10% detran sulfate. After incubation, unannealed probe is washed off and the slides are labelled with immuno-fluorescent tags (ie. antibodies to the labeled nucleotides carrying fluorescent tags) and an anti-fading agent.

In summary, the chromosomes have been exposed to methanol, acetic acid, flattening, drying, rehydration, heat, formamide, and high salt concentrations during the FISH procedure. Some degree of structural changes will have been produced in the chromosomes. High resolution microscopic equipment is required for FISH as well as image processing software to enhance and analyse the image.

FISH is being used to study chromosome banding patterns, the distribution of chromatin in interphase nuclei, and chromosome abnormalities. Greater resolution is possible in interphase chromosomes due to the dispersed state of the chromatin, so that small structural changes can be detected. The compaction of metaphase chromosomes sets the limit of detection at sequences separated by more than 1 Mbp. The limit in interphase chromosomes is a separation of 100 kbp.

Detection of Single Genes

When the probe is a single gene, only that gene is expected to light up in FISH. The hybridization of a probe for the *neu* proto-oncogene in hyman lymphocytes. The *neu* locus resides on chromosome 17. Since the cells contain diploid nuclei, two hybridizing spots are seen in each nucleus. The sensitivity of FISH is put into perspecive by considering the fact that each chromosome is a single DNA molecule. That is, each spot results from a few probe molecules hybridizing to different parts of a single DNA molecule. Looked at another way, we can recall that Avogadro's number (the number of molecules in a mole) is 6.02×10^{23} molecules/mole, we are detecting 1.66×10^{-24} moles of chromosomal DNA!

Chromosome Painting

It is possible to isolate DNA specifically from individual chromosomes. When total chromosomal DNA is used as a probe, it will base pair with all sequences on the chromosome from which it was derived, making the entire chromosome 'light up'. In other words, each sequence in the probe eventually encounters its complementary sequence on the chromosome and base-pairs with it.

STRUCTURAL BASIS OF CHROMOSOME BANDING

The exact role of underlying structural, functional and molecular organization of the chromosome in the determination of bands is still not fully understood. C-bands represent constitutive heterochromatin. G-banding patterns on mitotic chromosomes correspond very closely to the chromomere patterns of meiotic chromosome bivalents at pachytene, that is, the chromatin that is densely-packed enough to see. Chromomeres appear to be the focus of chromatin condensation along the chromosome, and may be the sites where chromatin condensation is initiated. If this is true, G- bands may be the corresponding initiation sites for chromosome condensation during mitosis.

Chromomeres may be more apparent in meiosis because of the relative degree of condensation of the chromosomes. Meiotic chromosomes are very extended and the homologues are paired which enhances chromomere patterns. Mitotic chromosomes are overall more condensed, masking chromomeres and requiring banding pretreatment to reveal them.

On human mitotic prophase chromosomes, up to 2000 or more bands can be observed. The bands on these high-resolution chromosomes correspond to meiosis chromomeres. As the chromosomes condense in the prophase to metaphase transition, there is a progressive coalescence of bands such that each band and its component sub-bands retain the same relative location and staining intensity. The bands observed at metaphase are made up of a collection of smaller sub-bands. The amount of constitutive heterochromatin in each chromosome is characteristic but polymorphisms do occur. The banding patterns are consistent and reproducible. Generally, features of chromosomes do not vary from tissue to tissue or change during the course of development. The organization reflected in banding is not random. In other words, we see consistent bands because the chromosome coils exactly the same way every time.

Constitutive heterochromatin is naturally differentiated by darker appearance during interphase and prophase, but does not usually show up in metaphase. It takes special methods to produce the bands. In plant mitotic chromosomes, C-banding has been produced, but not G-banding, although pachytene chromosomes of plant meiotic cells demonstrate chromomeres. One suggestion is that this is due to the much greater quantity of DNA per unit length in plant chromosomes compared to those in mammals. The chromomeres are too tightly packed to be resolved by G-banding. The reason is not clear as compaction seems to be roughly equivalent. The preparation or pretreatment may have to be adjusted.

PHYSIOLOGICAL AND MOLECULAR WHEAT BREEDING

Plant Breeding is art and science of improving the heredity of plants for benefit of mankind. In Evolutionary concept, plant breeding is merely a continuation of the natural evolution of the crop species, changing its course of direction in the benefit of greater use to mankind. This also be defined as science of selection-Plant breeding is essentially an election made by man of the best plants within a variable population as a potential cultivar. In other words plant breeding is a 'selection' made possible by the existence of variability. Selections become the earliest form of plant breeding.

History of Plant Breeding

Plant breeding started with sedimentary agriculture and domestication of the first agriculture plants, the cereals were chosen by the early man. They learned to look for superior plants to harvest. The domestication was hastened by early practice of harvesting mutant plants with special traits. This forms the ancient type of plant breeding. Before Mendal's discovery there was some plant breeding include selection and hybridization experiments. But the plant breeding was only hastened after discovery of Mendal's law on pea, thus lead a new science "Genetics". Modern plant breeding is applied genetics but its scientific basis is broader and uses conceptual and technical tools, molecular biology, cytology, systemetics, physiology, pathology, entomology, chemistry, and statistics (biometrics) and has also developed own technology.

Plant Breeding efforts may be divided into different historical landmarks-

Prehistoric Plant Breeding-Domestication of Crops

This includes domestication of crops by ancient people. Domestication is also continued so far.

Pre-mendel Plant Breeding

Economic botany and great Columbian exchange- There were some experiments before Mendel on hybridization and selection. Some seed companies were also established based on success on selection. Another part discovery of North America by Columbus in 1492 triggered unprecedented transfer of plant resources, first from old world to the new world, or the New World to the old. This increased variability in total genetic resources.

Mendelian Genetics and the Green Revolution

Mendel's experiment stimulated research by many plant scientists dedicated in improving crop production (plant breeders) through plant breeding. The most famous contribution of Mendelian genetics was hybridization. There was remarkable improvement in three economically important crops that made the food deficit world into a food surplus world. This is called the green revolution. The first, development of hybrid maize, the second development of high yielding and input responsive "semi-dwarf wheat" (CIMMYT breeder N.E. Borlaug received Nobel prize for peace in 1970), the third is high yielding "short sutured rice" cultivars. Similarly the remarkable improvements were done in other crops like sorghum and alfalfa.

Molecular Genetics and Bio-revolution

Totipotency shown by plants gave rise to tissue culture techniques such as somatic hybridization, doubled haploid production, clonal propagation, and in-vitro selection. Intensive research in molecular genetics has led to development of recombinant DNA technology (popularly called Genetic Engineering). Advancement in Biotechnological techniques has opened many possibilities for breeding crops. Thus mendelian genetics allowed plant breeders to perform some genetic transformation in few crops, molecular genetics provides key not only the manipulation of the internal structure but also their "crafting" according to plan.

Physiological and Molecular Wheat Breeding

Plant breeding has traditionally applied a trial-and-error approach in which large numbers of crosses are made from many sources of parental germplasm. Progenies are evaluated for characters of direct economic interest (*e.g.*, grain yield and grain quality) in target environments. Good performing parental germplasm, crosses, and progenies are selected for further use or testing. In many programmes "breakthroughs" in improvement are made simply by finding superior sources of parental germplasm among the numerous sources tested.

This conceptually simple approach has been highly successful in many crop species and numerous breeding programmes. The approach has often succeeded in the absence of in-depth knowledge about the physiological basis for superior performance. In some crops such knowledge has been obtained by doing retrospective analyses of prior genetic gains. Breeders have not applied this knowledge to a significant extent as a guide to further improvements, but instead have taken any avenue of improvement that happens to arise from direct selection for yield and economic performance. However with increased population, there is need to increase yield further and breeding require more scientific approaches to handle the problem.

Genetic Basis of Physiological Traits

During the past two decades, molecular tools have aided tremendously in the identification, mapping, and isolation of genes in a wide range of crop species. The vast knowledge generated through the application of molecular markers has enabled scientists to analyze the plant genome and have better insight as to how genes and pathways controlling important biochemical and physiological parameters are regulated. Three areas of biotechnology have had significant impact: the application of molecular markers, tissue culture, and incorporation of genes via plant transformation. Molecular markers have enabled the identification of genes or genomic regions associated with the expression of qualitative and quantitative traits and made manipulating genomic regions feasible through marker assisted selection. Molecular marker applications have also helped us understand the physiological parameters controlling plant responses to biotic and abiotic stress or, more generally, those involved in plant development.

Molecular Wheat Breeding.

Molecular wheat breeding is application of biotechnological tools in wheat improvement such as gene transfor (genetic engineering) and marker assisted selection. Such changes aims to alter the physiological pathways through change in genetic structures. There are many successful examples of such kind.

CYTOGENETIC TECHNIQUES

CHROMOSOME BANDING

A. Local variations in chromatin coiling, and the consequent variations in chromatin density, result in distinct staining patterns that can be used to identify each chromosome.

The discovery of techniques which produce banding patterns on chromosomes is one of the most significant developments in cytogenetics. These techniques told us that chromosomes have an identity and structure that is consistent in all cells within a species, supporting the chromosome theory of inheritance.

Staining made it possible to construct the first chromosome maps, and to distinguish between chromosomes of many different species. Fixed chromosomes are not equivalent to chromosomes in vivo. Fixation does not appear to extract DNA from chromosomes but it does extract histones and Non-histone proteins in varying amounts. The fixative may not always have a quantitative, reproducible effect on these proteins. The net result is that when you look at fixed chromosomes, you are looking at chromosomes that are chemically and structurally different from chromosomes in living cells.

Some relevant definitions:

- band - part of a chromosome that is clearly distinguishable from its adjacent segments by appearing lighter or darker.
- chromomere - any part of a chromosome that stains uniquely compared most bands.
- chromatin - the DNA/protein complex making up chromosomes
- heterochromatin - chromatin that tends to take up large amounts of stain, thus staining darkly. Heterochromatin tends to be densely-coiled.
- euchromatin - chromatin that tends to take up small amounts of stain, thus staining lightly.
- constitutive heterochromatin - chromatin that remains densely-coiled through out interphase, in all tissues. Often found near centromeres, telomeres and nucleolar organizing regions (NORs).

- facultative heterochromatin - chromatin that is either densely-coiled or loosely coiled during interphase, depending of the developmental state or tissue.

Giemsa Staining Methods

Mitotic metaphase is the best stage for studying chromosome morphology. Giemsa stain is specific for constitutive heterochromatin adjacent to centromeres and telomeres.

Geimsa C-banding technique has been used to identify individual chromosomes in many species by showing the position of constitutive heterochromatin.

Geimsa staining methods were first described in a paper by Purdue and Gall (1970). However, it is important to recognize that staining methods typically need to be fine-tuned for each species, to give the optimal number of bands.We will concentrate on the general treatments and the basis for their application in terms of what each treatment is doing to produce the banding pattern.

Steps of C-banding:

1. Roots are harvested, pretreated and fixed in 3:1 95% ethanol:glacial acetic acid for at least 24 h. The roots are softened in 45% acetic acid or in 0.5% aceto-carmine. Slides are prepared by the squash method and the cover glass is removed using dry-ice method. Chromosomes adhere to the slide surface.
   The acid pretreatment depurinates the DNA (ie. removes purine bases without cleaving the sugar-phosphate backbone. This helps to induce nicks in the DNA backbone, loostening any supercoiling.
2. Dehydration. Typically slides are placed in 95 to 100% ethanol for 1 hour.
3. Denaturation. Treatment with barium hydroxide for 5 to 15 min at elevated temperature 50-55° C to denature the DNA. Alkali treatment denatures the DNA. ie. breaks base pairing, resulting in ssDNA.
4. Renaturation. The slides are then washed with distilled water and transferred to incubation at 60° C in saline sodium-citrate solution SSC (NaCl) which presumably renatures the DNA. Incubation periods and temperature are variable.
   The incubation in SSC alters the composition of the chromatin (DNA-protein complex). The differential reannealing of the C-band and the non-C-band chromatin to extraction is the basis for the differential staining. The heterochromatin which is highly repetitive DNA renatures under these conditions while the middle repetitive DNA and unique DNA does not. The result is the C-banding pattern.
5. Staining. Slides are then stained with Geimsa stain and checked periodically to see how the stain is progressing. When the optimal

staining has been achieved, the slides are rinsed in distilled water to remove the excess stain, air-dried, stored in xylene overnight, air dried again and the cover slip is mounted using Canada Balsam, etc.

Other Geimsa banding techniques include G banding which provides more detail than C-banding. G banding provides identification of almost every chromosome in a complement and structural variation can be detected.

G-banding is produced using Giemsa (=G) staining and usually pretreatment with a diluted trypsin solution, urea or protease.

The Human Chromosome Study Group has published a diagrammatic representation of chromosome bands observed with G-,Q-, and R-staining methods. The nature of these bands appears to be stronger chromosome condensation, but the bands could also be interpreted as the result of alteration of histones and other proteins of the chromosomes. G bands cannot be produced in plant chromosomes possibly because of the high degree of condensation in plant metaphase chromosomes. N-banding was originally developed to stain the nucleolar organiser region of the chromosomes, plant and mammals.

CONCLUSION

Soon after the rediscovery of Mendel's Laws in the year 1900, the formulation of 'chromosome theory of inheritance' laid the foundation of cytogenetics, which has thus completed its first hundred glorious years in the year 2003. Subsequently, in the last few decades of the last century, the interest in plant cytogenetics that involved study and use of structural and numerical changes of chromosomes had largely declined. However, more recently, with interest in construction of molecular maps of chromosomes, and in the sequencing of whole genomes comprising an entire set of chromosomes, plant cytogenetics perhaps had a re-birth.

Other areas of recent research in plant cytogenetics include study of the organization of individual chromosomes in the form of chromosome territories and location of replication and transcription territories within the nucleus.

Chromosome structure has also been studied in great detail, so that the chromatin is now known to consist of nucleosome and chromatosome subunits, which undergo several levels of folding to contain it within the boundaries of the chromosomes. The structure of these nucleosome subunits has been studied at atomic resolution, and the structural relationship between adjacent nucleosomes has been recently resolved through X-ray crystallography of tetanucleosomes.

The structure of telomeres and centromeres have also been studied in several plant systems (particularly the cereals) in great detail suggesting that while structure of telomeres is conserved, but that of centromeres exhibit sufficient diversity.

In recent years, it has also been shown that both DNA and histone proteins, as components of chromosomes undergo large-scale modifications, which regulate gene expression. The study of these modifications also gave birth to the science of 'epigenetics and epigenomics'.

Projects like whole genome/epigenome sequencing of important plant systems are also yielding new information, giving new directions to research in plant cytogenetics. These exciting new developments in the field of plant cytogenetics will be briefly discussed in this discussion.

Bibliography

A.K. Zingare: *Plant Breeding and Seed Saving*, Satyam Publishers, Delhi, 2013.

A.R. Dabholkar: *General Plant Breeding*, Concept Publication, Delhi, 2006.

Bahar A. Siddiqui and Samiullah Khan: *Plant Breeding Advances and in vitro Culture*, CBS Publication, Delhi, 1997.

Denis J. Murphy: *Plant Breeding and Biotechnology : Societal Context and the Future of Agriculture*, Cambridge University Press, New York, 2013.

Dewasish Choudhary: *Plant Breeding*, Anmol Publication, Delhi, 2008.

Dharamvir Hota: *Modern Biotechnology in Plant Breeding*, Gene Tech Books, Delhi, 2007.

E.B. Babcock: *Genetics and Plant Breeding*, Agrobios Publication, Delhi, 2013.

Herbert Kendall Hayes, Forrest Rhinehart Immer and David Clyde Smith: *Methods of Plant Breeding*, Biotech Books, Delhi, 2009.

I.D. Tyagi: *Plant Breeding and Genetics at a Glance*, South Asian Publication, Delhi, 2005.

Iqbal Hussain: *Fundamentals of Plant Breeding*, Oxford Book Company, Delhi, 2009.

Iqbal Hussain: *Plant Breeding*, Oxford Book Company, Delhi, 2008.

James A.S. Watson, Wattie J. West and C.V. Dadd: *Crops: Varieties and Plant Breeding*, Satish Serial Publication, Delhi, 2005.

K.V. Mohanan: *Essentials Of Plant Breeding*, PHI Learning, Delhi, 1999.

Kendall R. Lamkey and Michael Lee: *Plant Breeding: The Arnel R. Hallauer International Symposium*, Blackwell Publication, New York, 2006.

Ludwig: *Dictionary of Plant Breeding*, Ivy Publication, Delhi, 2008.

M.P. Singh and Sunil Kumar: *Genetics and Plant Breeding, Vol. I and II,* APH Publication, Delhi, 2009.

Mallikarjuna Reddy and Aparna Rao: *Plant Breeding in Horticulture*, Pacific Books International, Delhi, 2010.

Manzoor Ahmad Khan: *Plant Breeding*, Biotech Books, Delhi, 2000.

Murli Krishnan: *Plant Breeding and Hybrid Seed Production*, Dominant Publishers, Delhi, 2012.

Neal C. Stoskopf, Dwight T. Tomes and B.R. Christie: *Plant Breeding : Theory and Practice*, Scientific Publication, Jaipur, 2006.

P C Trivedi and Indu Rani Sharma: *Cell Biology, Genetics and Plant Breeding*, Indus Valley Publication, Delhi, 2006.

P Gomathinayagam; K Vanangamudi; M Subramanian; K Natarajan and K R V Sathya Sheela: *Lexicon of Plant Breeding and Genetics*, International Book Distributors, Delhi, 2008.

R.L. Kapoor and M.L. Saini: *Plant Breeding and Crop Improvement (2 Vol-Set)*, CBS Publication, Delhi, 1997.

S. Dana: *Plant Breeding*, Naya Udyog Publication, Delhi, 2001.

S. Thirugnanakumar, K. Manivannan, M. Prakash, R. Narasimman and Y. Anitha Vasline: *In Vitro Plant Breeding*, Agrobios Publication, Delhi, 2009.

S. Thirugnanakumar, K. Manivannan, S. Manickavasagam and R. Ushakumari: *Organelle Genomes in Plant Breeding*, Agrobios Publication, Delhi, 2008.

S.K. Gupta: *Plant Breeding : Theory and Techniques*, Agrobios Publication, Delhi, 2008.

S.K. Kataria: *Plant Breeding : Theory and Techniques*, Educational Publishers, Delhi, 2011.

Sanjay Kumar Singh: *Plant Breeding*, Campus Books, Delhi, 2005.

Veenu Agarwal: *Genetics and Plant Breeding*, Oxford Book Company, Delhi, 2012.

Virendra Batra: *Genetics and Plant Breeding*, Oxford Book Company, Delhi, 2009.

Yasin Jeshima Khan, Srivignesh Sundaresan. P. T. Prathima, J. Nanjundan and M. Arumugam Pillai: *Merging Plant Breeding With Crop Biotechnology*, New India Publishing Agency, Delhi, 2012.

Index